Salt-Glaze
CERAMICS
An International Perspective

Bente Hansen, *Salt-glazed Vessels*, slip decoration.

Salt-Glaze
CERAMICS
An International Perspective

Janet Mansfield

CRAFTSMAN HOUSE

SALT-GLAZE CERAMICS: *An International Perspective*

First published by Craftsman House BVI Ltd, Tortola, BVI
Distributed in Australia by Craftsman House,
20 Barcoo Street, Roseville East, NSW 2069, Australia

Distributed internationally through the following offices:

USA	UK	ASIA
STBS Ltd.	STBS Ltd.	STBS (Singapore) Pte Ltd
PO Box 786	5th Floor, Reading Bridge House	Kent Ridge PO Box 1180
Cooper Station	Reading Bridge Approach	PO Box 1180
New York	Reading RG1 8PP	Singapore 9111
NY 10276	England	Republic of Singapore

ISBN 976 8097 11 6

Typesetting, Layout and Design   *Netan Pty Limited, Sydney*
Printer   *Kyodo, Singapore*

CONTENTS

ACKNOWLEDGEMENTS

I thank all the ceramists who have so generously contributed to the text and the illustrations for this publication and I trust it will be rewarding to them in the furtherance of their work. For information on the history of salt-glaze in Germany I am indebted to Wolf Matthes of Höhr Grenzhausen, and for the information on the salt-glaze potteries of La Borne I am indebted to Christine Pedley and Eva De Belder of La Borne, France.

1.

AN INTRODUCTION TO SALT-GLAZED CERAMICS

Within the broad field of ceramics, the art and technique of forming and firing clay for personal and cultural expression, salt-glaze holds a singular place. Salt-glaze, the result of a reaction between common salt and clay, gives a unique finish that forms an integral part of the ceramic object itself. The actual process of salt-glaze demands from its practitioners a full commitment to every aspect of the making, decorating and firing that is more intense than required by any other area of ceramic art. These processes allow spontaneity and direct action on the part of the ceramist and, with the many variations possible, salt-glaze practice can become a complex and unending source of experimentation and creativity.

Many contemporary ceramists are finding a potential for individual expression within the salt-glaze field, either drawing on tradition, or challenging its influence. These makers are concerned with pots, with domestic ware, with ceramic sculpture, with contemporary issues, or with the fulfilment of research and development. The ceramists described in this book are working throughout the world, absorbed in the process of salt-glazing and, despite all the difficulties associated with this technique, feeling that it allows them artistic involvement with materials and methodology, aesthetic judgement and a personal mode of expression.

This book will deal with ceramists currently using salt-glaze, concentrating on their motivation and the development of their ideas. Though their intentions may overlap the chapter headings of this book, they have been placed where their contribution to the salt-glaze aesthetic seems greatest.

An explanation of the technical aspects of salt-glaze is needed to introduce the reader to the process. This explanation will be developed further by the ceramists as they describe their work. Salt fumes have a dramatic effect on clay under heat. This occurs at temperatures from about 900°C, the melting point of common salt, when a surface blush of colour is formed on the clays and clay slips used by the potter, to over 1300°C, the traditional temperature for high-fire salt-glaze. At higher temperatures the salt becomes an active vapour; a typical salt-glaze has a glossy orange-peel texture enhancing the natural colour of the clay beneath it.

Common salt is composed of sodium and chlorine; these are both chemically reactive elements which combine with other elements, either in the clay or the atmosphere of the kiln. Sodium when thrown into a hot kiln acts as a flux or melting agent and reacts with the silica in clay to form a soda glass or glaze layer on the surface of the clay. To achieve a high gloss orange-peel salt-glaze, the clay is fired to its

(*Above*) Siegburg, *Schnelle*, mid-16th century, 24 cm/h, Keramikmuseum Westerwald, Höhr Grenzhausen

(*Facing page*) *Bartmannskrug*, Cologne, ca 1530, 31 cm/h, Kunstgenerbemuseum, Cologne

(Above) Pavement, La Borne

vitrification temperature. Different clays, depending on whether they have more or less silica in relation to the alumina in their composition, react differently to the salting; by a process of experimentation each individual potter seeks to find a clay which gives the desired surface.

More silica in a clay body will give a higher shine, while an excess of alumina will give a matt surface or low sheen. The amount of iron oxide present relative to the silica in the clay will be important in the final surface colour and texture. Salt fumes accentuate the iron component in the clay resulting in dull or shiny surface textures depending on the relative amounts of iron and silica present; greater amounts of iron will give dark brown to black colours while lesser amounts give tan or golden shades to grey. Colour is also dependent on the atmosphere in the kiln during the firing and the cooling rate of the kiln. These are some of the variables that can result in quite different effects.

Particular qualities of colour and shine can be exploited for surface variation using clay slips, oxides or stains. In addition, the particle size of the clay components, whether coarse or fine, will have a bearing on the final surface. The addition of coarse particle sands in the clay will result in a heavy texture with mottled colour, the coloured pigments melting off the high points and running into the hollows of the textured surface; finely ground ingredients will result in a smooth and evenly distributed gloss layer. The amount of salt thrown into the kiln, and the ease of passage of the salt vapours through the kiln, will dictate a light or heavy glaze layer.

Some ceramists like to remove draw trials or test rings, during the salting process.

(*Above*) Bottles, Westerwald (Grenzau), early-19th century, 22 cm/h
Collection: Wolf Matthes

(*Left*) John Dermer, vase, salt-glazed stoneware, cobalt and rutile slips, 1980

These are small clay rings made from the same clay body as the wares to be fired and usually placed alongside the set of temperature measuring cones. The rings are hooked out during the salting process and examined to discover the build up of salt on the clay body; they will also give some indication of the atmosphere in the kiln. The colour of the drawn ring will be the same as ware from a fast cooled kiln — usually grey; brown colours develop during a slower cooling.

The unique qualities of salt-glazing, with the glaze being formed from the clay itself, enhance the natural colours and textural quality of the clay. The ceramist is able to exploit every mark and gesture of the forming process, knowing it will be accentuated in the firing. Decisions, therefore, have to be made: from the choice of clay, decorative materials and methods, and the techniques of making, to the design of the kiln, its firing and cooling. At every step along the way the ceramist must make aesthetic judgement. Work that comes from the kiln is the result of the ceramist's skill and understanding in using processes that are not always predictable or easily controlled.

A brief outline of the history of salt-glaze is of interest. The value of salt-glaze was

(Right) Crucifix, salt-glazed stoneware, La Borne

of practical importance to society at the time of its discovery about five centuries ago. The possibility of making wares that were vitrified and glazed in one process offered an economy over the double process of bisque and then gloss firing. The salt-glazed wares did not leak, were impervious to acidic liquids and were strong enough to withstand daily use without readily chipping or breaking.

The technique of using common salt to provide a glaze is thought to have been discovered in the Rhineland in Germany where, due to the geology and climate, there are deposits of stoneware clays especially suitable for high-temperature salt-glaze. The discovery of the superiority of these clays was associated with technological advances enabling kilns capable of firing to high temperatures to be built. It is

possible that potters threw cooking salt into the flames and discovered by accident how to produce more refined glazed wares.

(Above) Janet Mansfield, jar, salt-glaze over slip decoration, 1300°C, 1979

In Germany, four distinct types of historically significant salt-glazed ware have been recognised: the light, almost white Siegburg stoneware; the brown engobe-coated wares of Cologne and Frechen; the grey, often blue painted stoneware from Raeren and the Westerwald; and the stoneware with yellow surfaces, often colourful and decorated, from northern Germany and from Saxony.

There were potteries in the area around Siegburg from the 12th century. Stoneware was developed in the late-13th or early-14th century and the glaze coating that can, with certainty, be described as intentional salt-glaze first appears in the 15th century. The technical high point of Siegburg stoneware occurred in the second half of the 16th century and the beginning of the 17th century, when richly decorated pieces in the taste of the Renaissance style were produced.

The brown Cologne and Frechen stonewares were mostly fired in reduction and, on cooling, re-oxidised on the surface resulting in warm yellow-brown to red-brown colours with a pronounced salt-glazed surface. It is especially thick and powerful on the so-called *getigerten* (tiger) ware, beloved as an import ware in England in the 17th century. The most typical example is the *Bearded Man Jug* from Frechen and the elegant oak bordered pieces from Cologne. Salt-glazed stoneware in the grey

(*Above*) End of the firing, St Armand
de Puisaye.
Photograph: Claude Presset

(*Right*) Tureen, Westerwald (Höhr
Grenzhausen), 20th century Werkstatt
Höfer, 27cm/h
Collection: Wolf Matthes

reduced form, often with cobalt blue painting, was produced in Raeren and is possibly an invention of that area. This grey reduction-fired stoneware, often with colourful cobalt smaltz decoration, is also typical of the Westerwald area in Germany and is still produced today in the vicinity of Höhr and Grenzhausen, the area known as *Kannenbäckerland* (Jug-baking land).

Each locale had its typical forms, colours and decorative style; fashion played its part and similarities of style were spread around by wandering potter journeymen and glass painters. During the Baroque period the painted decoration with enamel colour overglaze on ware from Creußen and Freiberg was popular. However the style of decoration using cobalt containing blue smaltz on grey reduced clay body became popular everywhere in the 18th, 19th and 20th centuries.

There are numbers of ceramists in Germany today using salt-glaze techniques in their work incorporating their own ideas of form and decoration. They continue to experiment and their work is influencing others. In this way they become part of a living tradition.

(Left) Jug, La Borne

The stoneware deposits of middle France in the old centres of Saint Amand en Puisaye and La Borne stimulated a similar flourishing of salt-glazed stoneware kilns and, as with the German potteries, met the needs of their communities in producing not only domestic pottery but also bricks, tiles and pipes.

La Borne is a small village in the centre of France, about 30 kilometres from Bourges. Here potters found excellent clay and an abundance of wood. At first their work was for local use, but gradually their production expanded and was exported by train and barges on the Loire river as far as Holland and overseas to America. The activity and production in this small village became immense.

(Right) M. Gaubier, St Armand de Puisaye,
1952
Photograph: Claude Presset

The old French kilns, lying on their sides in the shape of half-buried flasks and resembling carp-backed Chinese kilns, measure 10 m long by 3.5 m wide, and have a capacity of up to 80 m³ with the firebox at the narrow end, or neck. With these extremely difficult and temperamental kilns, the potters achieved stoneware firing temperatures of 1250 to 1300°C. The kilns were packed without shelves, the pots stacked on top of each other. The insides of the pots were glazed with ash glazes, the outsides left unglazed.

The clay in La Borne is rich with minerals, almost black when soft, grey when dry. The combination of flame, ash deposit and salt vapour from small goblets of salt placed throughout the kiln gives it subtle shades of colour and a satin-like surface quality. It is said that a La Borne pot always has two sides: a face and a back. The face is where the flames reach easily and the pot receives an ash deposit, and the back is where the pot has been shielded.

(*Above*) Milk jug, La Borne

(*Left*) Jeff Mincham, jar, ash glaze under salt-glaze, drip from roof, cone 9, 49 cm/h, 1979

As the potters did not use shelves they placed the salt goblets on the handles of the big jars with small pieces of clay. In this way the salt vapour was spread through the kiln. The packing of the kiln took several days and the firing itself went on for at least five days, depending on the size of the kiln, the quality of the wood, the weather and the moon. It was common belief that firing during the period of the rising moon helped the kiln reach high temperatures. One knew the firing was coming to an end by the acid smell and the heavy black smoke of a reduced firing.

The story of salt-glaze will be found wherever there was clay to make the wares, wood or coal to fire it, and the technology to build kilns that could fire to stoneware temperatures. From Germany and France the production of salt-glazed wares spread to other European centres and to England, where the vast deposits of ball clays in the south-west of that country were most suitable for salt-glazed stoneware. The peak of English salt-glazed wares was at the end of the 18th century, although John Dwight gained a patent to produce salt-glazed stoneware in 1671. He was followed by the Elers brothers and, in the area around Staffordshire, the making of salt-glazed wares flourished. Doulton wares from Lambeth, in particular their art ware pieces, became popular in the 19th century. These were extravagant forms, appropriately decorated, with each part of the process being undertaken by specialist craftspeople.

(Right) Don Reitz, loading the salt kiln,
Wisconsin, 1977

Many studio potters in England today are interested in salt-glaze, and keep the
tradition alive while making work that is both contemporary and personal, undertaking all the aspects of the craft themselves. Bernard Leach, who used salt-glazing
techniques in his own work, gave good descriptions of the process in *A Potter's Book*
(Faber and Faber, London, 1940).

Colonial wares, first in the USA and then Australia, were salt-glazed as soon as
clays were discovered and there were immigrant pioneers with the skills to build the
kilns and fire the work. With the use of lead in glazes being recognised as toxic in the
19th century, many potteries in the United States decided that salt-glazed stoneware
would be a safe alternative. Early colonial potteries making salt-glazed stoneware
were established in New Jersey, New York and Philadelphia and they provided
strong and vitrified jars needed to store and preserve food. Potteries were started in

New England, then in the south and the mid-west and later in the western States, making brown and cobalt-decorated grey salt-glazed jugs, crocks, bottles and churns, now much sought after by collectors.

The current studio pottery interest in the USA in salt-glaze has largely been as a result of the work of Don Reitz who investigated the process as part of his graduate degree at Alfred University in 1959. Don Reitz's ability as a ceramic artist and as a teacher has influenced many others to experiment with the technique in their own work.

In Australia, potteries were established in the early settlements of Lithgow, Bendigo, Adelaide and Brisbane, using local clays and coal for fuel. Salt-glazed wares for practical purposes were handsome in form and surface. As well as pottery, they made such items as sheep troughs, chimney tops, pipes and tiles for the growing colony. Artist potters used the techniques from the 1920s and Carl McConnell used salt-glaze in his repertoire of high-fired stoneware in his Brisbane studio in the 1950s. By 1975 a number of Australian studio potters began to explore the process. Jeff Mincham from South Australia, John Dermer in Victoria and Janet Mansfield in New South Wales were among those potters that had successful exhibitions of salt-glazed ceramics in the 1970s.

Salt-glazing in Japan has a short history relative to the thousands of years of ceramic tradition in that country. The Idemitsu collection holds one Bellamine-style jar, made in the 19th century at the San Kama kiln, which was obviously inspired by German models, and the Tokoname Institute of Ceramics also holds some early pieces. It was perhaps Hamada Shoji's interest in salt-glaze in the 1950s which encouraged others in the technique, with the result that it has become more familiar and practised in Japan.

The story of salt-glaze, which continues to evolve world-wide, is taken up in the chapters which follow, for salt is the common factor used by all the ceramists represented. Their intentions, materials and techniques are examined using their own words, the words of critics, and the author's comments. This book is not intended to be a technical manual, however artistic intent and technique are closely related in ceramics and are therefore part of the total expression. Each artist's work is discussed as a complete and separate story. The cross-referenced index will help in the search for particular technical details such as the formulation of clay bodies, decorative styles and firing schedules.

To date, the few publications on contemporary salt-glaze practice included Don Reitz's monograph written in collaboration with others for the American Craft Council (produced in conjunction with an exhibition in 1972), and Jack Troy's book, *Salt-Glaze Ceramics*, published in 1977. Peter Starkey's text on salt-glaze in the Pitman Skill Books series was also a welcome addition to contemporary documentation of this technique. A list of reference texts for further reading may be be found in the final pages.

I hope this book will encourage ceramists and collectors alike to search out the potential of this specialty area of the ceramic arts. The rewards are there as the following chapters will show.

(*Above*) Bottle, screw topped, enamel painting, Creusseu, 1680, 37.5 cm/h Collection: Hetjens-Museum, Düsseldorf

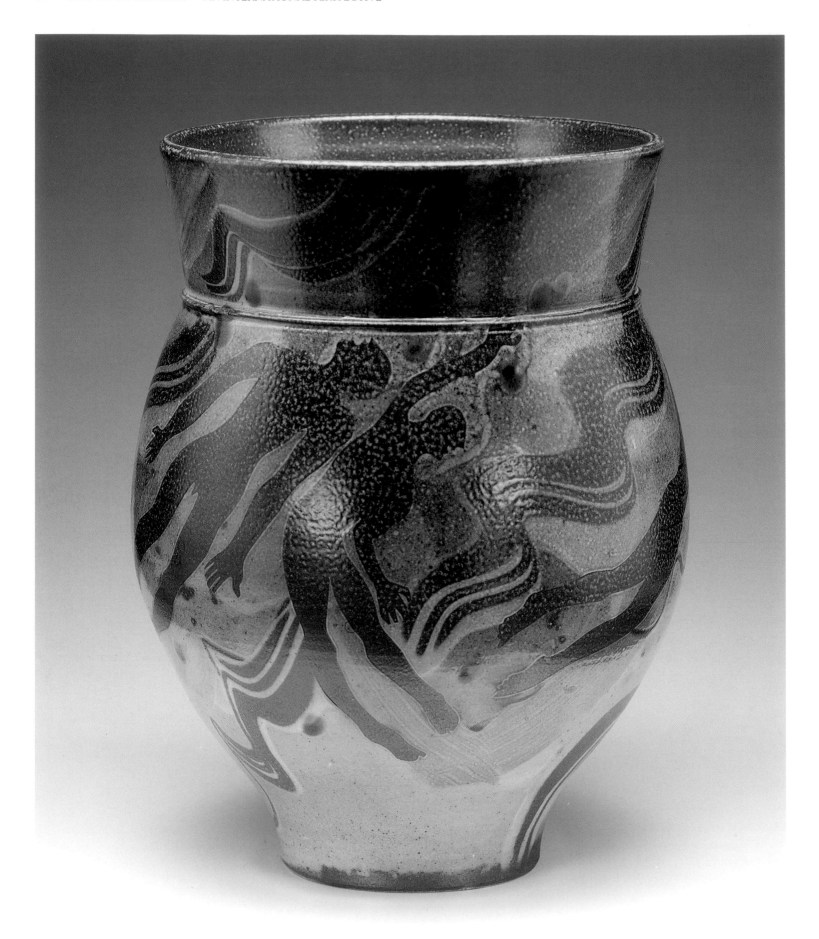

2.

A CELEBRATION OF POTTERY

There are some potters who, by their work and attitude to the art of pottery, draw attention to all that is good about making pots; the joy of making, and the possibility of expressing ideas through this art form are evident. One can see in the use of the clay and the fire their certainty and familiarity with the techniques and their love of making pots. Such potters are portrayed in this chapter.

Michael Casson's pots have met with acclaim since his first exhibition in 1959. Since that time he has been exhibiting and teaching around the world and has written a number of publications on ceramics. He set up his present workshop, at Wobage Farm, Herefordshire, UK, with his wife Sheila, also a potter, in 1977.

'All my works are functional,' he says. 'During the last fifteen years or so I have tended to make vessels for occasional rather than everyday use, but they are all still for holding liquid, cooking, serving or storing food, brewing tea etc. It is only in the last 10 to 12 years that I have turned to wood-fired salt-glazed stoneware. The directness of the technique attracted me as much as the actual surface qualities it gives. I have always believed that the insight that comes from the understanding of the materials and processes can, in itself, release the imagination to become a powerful stimulus to the creative process. Of course, the vision is needed, but its source may vary — at times it comes from the clay or the fire itself.'

Writing for *Crafts* magazine in 1989 in an article titled 'Sources of Inspiration', Tanya Harrod further emphasised Michael Casson's stand. She quoted him: 'Materials and processes are as vital to the imagination or to the stimulation of the imagination as any more conceptual source material'. Harrod continued: 'Education, Casson believes, should illuminate the materials, the processes, the sheer physicality of making pottery. The process of the making, the perils of firing, the heat and the stress should serve to ignite, metaphorically, the imagination; these processes are part of the narrative content of pottery. For Casson, the medium is the message. And as time has gone on, he has ceased to concentrate on domestic ware but rather on certain key forms. Take his majestic jugs for instance. Nothing could look more reassuring and potentially useful. But in fact they are too large to serve as pouring vessels. They stand rather for all the qualities of jugs, born of looking at thousands of other jugs made in our ceramic past. In that way the conceptual side of Casson's art is firmly rooted in the ceramic process and its history'.

Michael Casson uses two clay bodies, one a Harry Davis recipe made from ball clay and porcelain powder china clay plus 1% Cornish stone, which makes a white

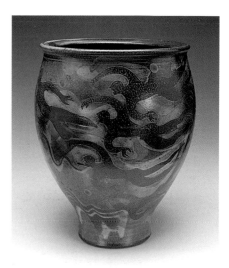

(Facing page and above) Michael Casson, *The Swimmers*, salt-glazed stoneware, resist decoration

(Right) Michael Casson

siliceous stoneware giving a smooth non-pitted surface and the other a mixture of china clays, fire clay, and a red earthenware clay with 6% of sand added. He says of his techniques: 'All the work is thrown and many pieces are made by joining a wet thrown coil to a dryish base form to give the final vessel. Various surfaces are obtained by dipping, brushing or pouring clay slips singly or in combination, layer on layer. Sponge, comb and finger wiping is used. Recently, the paper resist method has been used to give the figurative decoration — used in a manner similar to printing: the colours are built up in layers with the paper resist used in between and on top of the different layers'.

Clay slips used by Michael Casson are mostly a mixture of nepheline syenite and ball clay (50/50) and coloured with cobalt, rutile or titanium to give blue, tan or

(Left) Michael Casson, jug, salt-glazed stoneware

cream-tan tonings. The tan slip is bright if well reduced and the cream-tan slip is lighter or darker according to the clay body used. Washes of slips or of red earthenware clay are applied thinly and affect the other colours; for example, the red wash will darken the other slips. All slips are applied to the raw pot before bone dry hardness is reached.

Glazes used as liners for the interiors of his pots are: *Tenmoku*: 75 Cornish stone, 15 whiting, 10 china clay, 4 crushed iron scale, 4 iron spangles; *White Shino*: 38 petalite, 38 nepheline syenite, 10 iron bearing AT ball clay, 14 ball clay hiplas (71); *Ash blue*: 40 potash felspar, 40 ash, 20 ball clay, sometimes with additions of talc and/or quartz depending on the ash, plus additions of 2% iron oxide and 1% cobalt oxide for colour; *Magnesium white*: 63 Cornish stone, 13 whiting, 7 ball clay, 13 china clay, 4 magnesium carbonate. All his firings are within the range of 1300°C, cone 10 flat.

Michael Casson's pots celebrate all that is robust and healthy. They look strong, with firm, clean lines, well-formed and rounded rims, and handles and lugs that can be grasped with confidence. The pleasure in the making of these pots is transferred to the pleasure in their use.

(Right) Sandra Johnstone, teapot, salt-glazed porcelain, hinged lid with brass pin, 20 cm/h

Sandra Johnstone of California, USA is another potter who makes pots that show the joy of the making. Her pots come from the wheel with seemingly effortless ease, and to see her throw a large ball of clay, opening it from the top, gradually working her way down the mass of clay before drawing up the sides, is to see how clay responds fluidly to the movements of the potter's hands.

With a Bachelor of Arts degree from Scripps College, Claremont, California under Paul Soldner, work as a graduate student at the University of California under Peter Voulkos, followed by a Master of Arts and a Master of Fine Arts from San Jose University under David Middlebrook, plus a teaching credential from the same institution, Sandra Johnstone is well qualified for her role as teacher and self-supporting studio potter.

She talks about her work: 'My direction throughout my development as a potter has been to simplify techniques and forms. I feel that I could have worked with some level of success in another firing technique, but high-fired salt continues to interest and challenge me. Often when unloading the kiln I see glaze results more appealing than anything I could plan or imagine'.

Johnstone began exploring salt-fired pottery in 1969. Before she built her first salt kiln she discussed it with her teacher and mentor, Paul Soldner, who told her not to

put a brick floor in her kiln, not to use kiln shelves and, instead of putting a damper in the chimney, to leave a few loose bricks at the bottom which could be removed to reduce the kiln atmosphere by inhibiting the draught in the chimney.

(Above) Sandra Johnstone, plates, salt-glaze over slip decoration, stacked in the kiln separated with clam shells, 20 cm/d

Johnstone built a 1.7m³ (60 cu.ft) salt-firing kiln at her studio in Los Altos, near San Francisco, California, from a catenary arch downdraft plan, and uses eight burners made from standard pipe fittings. The kiln has a floor of earth covered with lime. The salt ports are in the front wall. The kiln is fired to cone 9, reduced during the salting, and stoked once with approximately 10 kg of salt. The inside of the kiln is painted with high alumina kiln wash and has been fired hundreds of times. The kiln is stacked with greenware only (one firing is traditional for salt-fired ware) and she uses no kiln shelves.

Of this she says: 'This means that the pieces are made with special stacking requirements in mind. The challenge is to make pieces which are strong, not too thin at the top but also not too thick at the bottom because the weight of a thick wall would cause more stress on the pots, especially those at the base of the stack. Though it was two years before I was convinced that the approach I had taken was reasonable, I believe that this requirement has caused me to make better pots. Pottery which is even-walled and designed for structural strength is also visually stronger'.

The artist mixes her clay body in her studio using California fireclay and sand, finding that the effect of salting on this clay mix is particularly effective. She uses a clear liner glaze on the interiors of the pieces, taking care that there is no glaze where pots, especially their rims, will come in contact with one another. Slip decoration is an important part of her work and slips are applied by pouring and brush-

(Right) Claude Varlan, jar, 50 cm/h

ing on areas of pots which will be exposed to the salt vapour. Because the pots are touching each other during the firing, the glaze deposited is not all the same thickness. The kiln adds variation to the surface with patterns of flashing and areas of thick and thin glaze.

She says: 'Stacking plates closely promotes vapour activity between them; this produces pink and orange blushes and a flashed halo occurs around the mark of the shells I use as separators, enriching the slip decoration. Another variation I have added in the past few years is the use of saggars. The buyers of my pots have requested more porcelain and, as the space for it in my kiln is limited by the fact the porcelain is so glassy that the pots fuse together when they are stacked touching each other, I throw short cylinders, about 35 cm in diameter, which have holes cut in them (like Swiss cheese). These are stacked vertically with each saggar holding about five pots. The saggars are dipped in kiln wash and they survive about six firings'.

Johnstone says she enjoys working with limitations, especially self-imposed ones such as her stacking methods, because she believes that they lead to invention which is otherwise unlikely. 'Stacking the salt kiln without shelves has led my work

(*Above*) Claude Varlan

(*Left*) Claude Varlan, jar, salt-glazed in wood kiln, 55 cm/h

to change in response to the process and has given it an individual quality. The problem of physically handling a large piece of clay so that the result is lively, fresh and spontaneous is a continuing fascination to me. I have not tired of the problem because my success with the solution to it has increased as I have learned more. I feel that my pots exemplify a kind of paradox. The clay which was once soft, plastic and responsive is altered through the firing process to a dense material, hard and irreversible. I want to bring the initial fluidity through the process so that my work has movement and spontaneity. Even the largest pieces, though massive in scale and structure, should retain the softness and movement of life.'

In addition to her regular teaching at San Jose University, Sandra Johnstone has given many workshops. Her work is shown at over forty outlets (established galleries and sales contracts) in the United States and has appeared in many juried and invitational exhibitions, and in publications on ceramics.

For Claude Varlan and Brigitte Pénicaud of Prissac, France, making pots seems a joyous and celebratory affair. Both have studied and worked in a number of potteries and had numerous exhibitions since 1979. They use clay from Saint Amand en

(*Above*) Brigitte Pénicaud, plate

Puisaye, and decorate it with slips and coloured clays, oxides and pigments, using fingers and brushes, 'all in the spirit of grand liberty'. Wood firing in a kiln with three chambers and a capacity of 9 m³ lasts 24 hours. Small pots containing salt are disposed throughout the kiln in order to obtain the desired effects on the pieces, which are piled up or laid on refractory wedges or props.

'The big pieces are made in two parts,' explains Claude Varlan, 'and assembled to obtain the desired form. I believe the essential thing of our work is that it is made with the greatest freedom possible. We are not interested in portraying sentiment in our work and we refuse to accept limitations'.

In an article titled 'Varlan — The Potter', Robert Deblander wrote that Varlan had 'learned without a break, not with the abstract and limited mind of a scientist, but that of a real human. One must find the work, experience it, feel it, sense it. Penicaud and Varlan,' he continued, 'are not in fashion, they are before and after. Today, or in a century's time, there are people who like water, rocks, trees and wind and objects that look like them'.

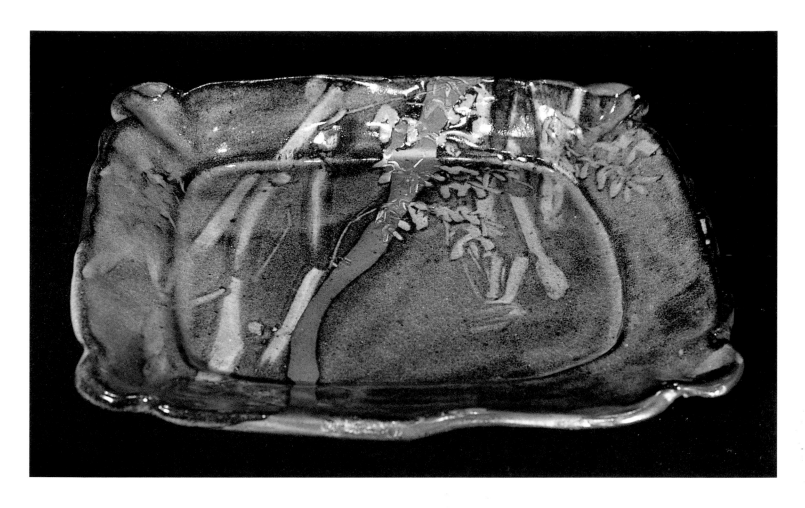

The ceramist Michel Gardelle, writing for *La Revue de la Ceramique et du Verre* in 1988, drew attention to their combination of talent and technique, enabling them to make pots that 'conceive generous volumes, are stable and with a beauty that opens the way to multiple visions of full shapes, sensual material and colours, tangled with the past. 'Their adventure,' he writes, 'gives the necessary strength to their pots'.

(*Above*) Brigitte Pénicaud, plate

Marielle Ernould-Gandouet, in *L'Oeil, Revue D'art*, April, 1989, wrote that Claude Varlan and Brigitte Pénicaud, in an exhibition at Galerie Daniel Sanver, Paris, were 'writing the language of pottery on volumes of grand earth. The slips, the fire, salt and a little time give the radiant colour to the glaze and never allow us to forget the presence of the flame'.

A similar celebration of the fire is part of the motivation of Patrick Sargent's work. From his early student days it has been a love of the firing process that has directed his style of work and his working life. Sargent is an English potter who has established his home and workshop in the lush Emmental hills of Nyffel, Huttwil in Switzerland. Here he has built a 5 m³ anagama-style climbing kiln firing up to 1350°C over two to three days. With a Bachelor of Arts with Honours in Ceramics from West Surrey College of Art and Design, Farnham, he has gained numerous awards for his work and held exhibitions; his teaching experience includes giving lectures and demonstrations and building kilns.

Sargent's particular specialty is wood-fired kiln design and construction, kiln theory, and firing techniques using the wood fuel as a creative and dynamic part of the ceramic process. He writes: 'My affair with pots and work as a potter has always been concerned with the roots, those simple beginnings and the essential elements

(Right) Patrick Sargent, vase, wood-fired
and salt-glazed, 26 cm/h

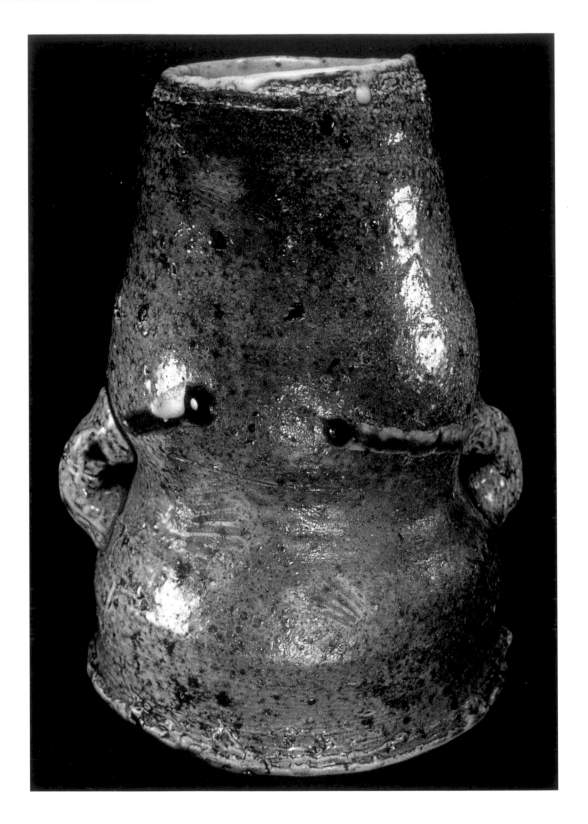

of the pot making process that seem to have virtually disappeared as a direct result of modern trends. There are few potters who understand real fire, the majority prefer-ring the lure of technological innovation and contemporary modality. Thus objects are being produced from clay which may well be technically and intellectually brilliant but sadly lack heartfelt aesthetic or loyalty to the origins of the craft.

'As a result my pot-making development has been inextricably bound up with kiln development. Not in a technological manner, but in an instinctive and feeling way. The nearer I can get to the process of nature in the way I make and fire, the clos-

(Left) Patrick Sargent, plate, wood-fired, brushed slip and stamped decoration, 32 cm/d

er I get to the roots. My salting technique is simple as I use salt as an enhancer. About two hours before the end of the firing I throw in 2 kg of salt at the front and the side stoke holes. The kiln dampers are opened to allow a rapid flash of salt throughout the kiln. Those pots nearest the openings become heavily salted.

'The kiln that I have now realises the dream of a decade. Its conception and construction had its beginnings at Farnham ten years ago. The pursuit of that rarely attained harmonic balance between clay, surface quality and fire is possibly my major motivation. The long fire contains an element that is as potentially creative as the initial lump of clay and that alone presents me with sufficient horizons for a lifetime's work.'

The potters who have contributed to this chapter have revealed an understanding of what is involved in making good pots. Their words communicate the excitement of continual discovery and their love of making pots, of using the materials and the processes, allied with their skills and intention to make something worthwhile, is evidence of their continuing celebration of pottery. When clay comes through the fire still retaining those clay-like qualities that inspired the potter in the first place, and when the processes of making and firing are there to give pleasure in the having, holding and using of these pots, then the potter can feel satisfied. There will always be a place for these potters in our contemporary society.

3.

THE DIGNITY OF DOMESTIC WARE

'There is a degree of dignity in being a domestic potter in the late twentieth century,' wrote Peter Dormer in a foreword to an exhibition of Jane Hamlyn's work at Leeds City Art Gallery (UK) in 1988. 'The dignity derives, in part, from the fact that a potter like Jane Hamlyn has control over her life.' This means that she is able to make aesthetic choices in regard to the materials she uses, her methodology and her working environment.

Making domestic ware can be a means of livelihood as well as a potter's love or leisure. This chapter will reflect the thoughts and challenges of a number of potters who make pots for use and who believe in and enjoy all the processes that salt-glazing entails. To be successful at an artistic level, however, requires more than enjoyment, belief and technical expertise.

Jane Hamlyn has this ability to bring domestic ware into the artistic domain. As W.A. Ismay commented in a review for *Crafts* magazine of Hamlyn's exhibition 'Use and Ornament' at Leeds: 'She has been devoted throughout her career to one particular discipline, that of high-fired salt-glaze. It is this kind of persistent dedication (given also the talent, the power of self-criticism, and the impulse always to reach out to something beyond) which is apt to produce better pots'.

Jane Hamlyn's contribution to contemporary ceramics shows that a lifestyle of dignity and self-determination is possible in making work that is sought after, relevant and exciting within the domestic ware idiom. Hamlyn studied at Harrow in the Studio Pottery Course, 1972-74, under Michael Casson and set up a workshop in 1975 in rural Nottinghamshire, UK, working with her husband, painter Ted Hamlyn. She is a visiting lecturer and external examiner at various art schools; she has given workshops, exhibited her work in many countries and her pottery is in public collections.

Hamlyn says of her work: 'I believe that useful pots have a unique role to play in the arena of creative art, in that they provoke audience participation. By handling a pot, considering how to use it, filling it with fruit, even enjoying the washing-up, the user engages in the creative life of the pot. I like that concept. In some ways, of course, it is madness to use salt-glaze for domestic ware especially if you hope to make a living by it. The range of tolerance, in terms of surface, allowed to useful pots is much less than that considered acceptable for sculptural work. Roughness, uneven salting, flyash, drips, and other vagaries of firing are much less obviously discernible on an art piece and can even enhance it. Conversely, these imperfections are usually

(Facing page) Arthur Rosser, jug, slip decoration, salt-glazed in wood kiln, cone 12, 35 cm/h

(*Above*) Jane Hamlyn, serving dish, cobalt slips and glazes, salt-glazed
Photograph: Bill Thomas

interpreted as denying, or at least limiting, the functionality of a piece which has been designed for use. *Seconds* debase the currency both in aesthetic and economic terms, but by destroying them in order to maintain quality the price of the remaining pots must rise. However, the current market is reluctant to accept the notion that useful pots deserve the same prices as useless ones.

'With salt-glaze you take a gamble, so much depends on getting the firing right and success is often hard won. There is the tedious slog of kiln maintenance and grinding alumina from lids and bases. Then, how much salt to use is a critical decision; you try to strike a balance between too little salt, which means unmelted surfaces and *underdone-* looking, boring pots, or too much salt with gross orange peel and glaze that runs down and sticks pots to shelves and bleaches out colours. It is important that the size of orange peel texture is in proportion to the pot, and in an attempt to control this, I use a sandy clay body for bigger pots and a fine white stoneware for smaller pieces.'

After raw glazing the interior of the pieces with a Shino-style liner glaze, Hamlyn paints the exterior, when bone dry, with a dark blue slip. This slip has 1.5% cobalt oxide and 1% titanium dioxide added to a base of 62 nepheline syenite and 38 mixed

ball clays. Over this slip layer, areas are brushed with a green glaze: 18 whiting, 26 potash felspar, 27 ball clay, 29 silica, 6 iron oxide. The piece has a final coating of a thin titanium wash to selected areas.

(Above) Jane Hamlyn, jugs and teapot, salt-glaze over slip decoration
Photograph: Bill Thomas

Hamlyn likes to use enough salt to ensure that the top layer of the pot is actually melted. This gives 'lush, rich, juicy-looking surfaces. I use slips and glazes which work with more or less salt, depending on their position in the kiln, that is, in relation to the charge of salt at the salt ports. However certain colours, for example an acid green lustre from titanium and cobalt, are very dependent on subtle nuances of salting (and reduction) which are largely out of my control'.

The artist continues to work in salt-glaze 'because it can be so beautiful, because it brings out such a wide spectrum of responses in the colours and surfaces of slips and glazes. It offers so much feedback in the form of unexpected results which suggest further possibilities for development, like a dialogue. I like pots which reveal the material of their making — clay; and the state of clay for which I have a particular fondness is soft, not wet, not sticky, but soft. I love the way that soft plastic clay is so responsive. Most of my handles are formed by pressing soft clay with or into various textures: rubber flooring, car mats, bits of wood etc. I also use roulettes and impressed stamps and some pots incorporate a wheel thrown and distorted ring joined to a slab-rolled base textured with embossed wallpaper. All these textures, which in them-

(*Top*) Mirek Smisek, crock, fluted decoration, salt-glaze over cobalt slips, 47 cm/h

(*Above*) Mirek Smisek, wine bottle, fluted, salt-glaze over manganese slip, 30 cm/h

selves come from mechanically designed found objects, are enlivened and varied by salt-glazing. I am interested in colour and texture, variety, excitement and unpredictability and in the balancing act between control and excess. These are the keys to my fascination with salt-glazing.

'I set things up and hope they will happen, although I can never be sure. To reveal the fine modelling and clarify the form I like a clean flame. When everything goes right, when heat and flame and salt combine to enhance what you have made, it is magic. Something extra happens, something beyond you — a revelation.'

Jane Hamlyn's work has a distinctive style. It is highly decorative, glossy and has a perfect finish. The richness and lustre of the colours and detailed patterning, impressed lines and embossed textures give the work a precious quality, reminiscent of ornate silver ware. Her jugs, serving dishes and drinking vessels seem to be designed for celebrating a special occasion.

One of New Zealand's pioneer and foremost potters is Mirek Smisek who began making pots in 1948. After working in ceramic factories in Australia and then in New Zealand where he settled, he decided to make the move to full-time studio pottery in 1956. Later he studied in Japan, and then spent a year at Bernard Leach's pottery in 1963. Here, he says, he 'began to learn what was really involved in making a handmade pot'. In a catalogue introduction to a 1983 exhibition by Smisek, writer Neil Rowe described the artist's year at St Ives with Bernard Leach: 'Leach's humility and intuition were paramount. Vigorous daily discussions with him, his generosity as a teacher and his deeply humane philosophy made a lasting impression'.

On Smisek's return to New Zealand, he built two kilns for salt-glazing, a medium to which he has been dedicated ever since. His studio at Te Horo in New Zealand's north island was established in 1968. Here, two beehive kilns for salt-glaze are visible from the road. Rowe commented: 'There is a joy and a zest for living inherent in every piece and his domestic ware, which provides lasting pleasure to the user, is the mainstay of his production'.

Sometimes, seeking more sculptural dimensions to his forms, Smisek alters the pieces when they are still soft, gently pushing them to give suggestions of movement and asymmetry. For additional texture on the external surfaces, he dips his wet hands into a mixture of fine granite and mica sands and then continues to throw. Smisek says of his work: 'A pot which is used several times a day puts a big demand on its maker. It needs deep involvement, feeling for materials, understanding of firing, sincere dedication and discipline. Motivation for every pot, be it a humble mug or a large vase, should show a genuine commitment and desire to create a pot which has warmth and vitality in generous measure.

'Creativity is one of the most important activities for people to engage in. My experience during the last war in my country, Czechoslovakia, more than convinced me of this. Many of the problems of the contemporary world will be minimised if creative activity becomes part of all our lives. Pottery, with the exciting challenge of mastery over the elements of earth, water and fire, offers tremendous scope for fulfilment. It is demanding, and good results do not come easily, but there is a great adventure for anybody willing to be sincerely involved. As a potter, my aim is to utilize and highlight the rich textures contained in our clays and rocks. I use clays from the earth near where I live and prepare them myself, and identify with their inherent qualities. It is important to aim to make pots which fulfil our needs and to surround ourselves with aesthetically healthy objects which should not only be admired for their beauty, but have much of their fulfilment in frequent handling and use.'

An apprenticeship at the Leach Pottery in St Ives, England has left its indelible mark on a number of potters. Mirek Smisek recalls Leach's generosity and total

commitment to making pottery that was both useful and beautiful. This is corroborated by Byron Temple, an American potter who, after studying ceramics at Ball State University, the Brooklyn Museum School and then at the Art Institute of Chicago, undertook an apprenticeship with Bernard Leach from 1958 to 1961.

(Above) Mirek Smisek, beehive kilns, Te Horo, New Zealand

Temple writes: 'Leach was always available, he would tell you all he knew'. The example of Leach, plus a strong interest in the clean-cut lines of Scandinavian design has resulted in a distinctive personal style. 'As a production potter I limit myself to designs which can be easily repeated. I do not find this restrictive or inhibiting, rather, I am able to explore more intensely the fundamental qualities of form, craftsmanship and expression.' Since 1962, Temple has run his own workshop at Lambertville, New Jersey, refining and simplifying his work to express his own values in his own way. He says: 'I strive for purity and precision in objects that extol the virtues of harmony and proportion'.

In an article on the work of Byron Temple for *Metropolis: The Architecture and Design Magazine of New York* in September 1982, Sarah Bodine wrote: 'Inspired by what he called the Sung standard, Leach advocated truth to materials and a striving toward unity, spontaneity and simplicity of form'. In Byron Temple's pots she found exemplary examples of this philosophy 'in the subordination of all attempts at technical cleverness to straightforward unselfconsciousness'.

Salt-glaze has been an important part of Temple's repertoire, suiting the practical nature and aesthetic worth of his production wares. The surfaces of his pieces are mostly undecorated, relying for their impact on the subtlety of the salt-glazed textural variations and the spontaneous and fluid character of his throwing.

Byron Temple's work has been exhibited in many countries and featured in many magazine articles. His teaching experience has included work at the Philadelphia

(Right) Byron Temple, tie box, salt-glazed,
12 cm/h, 1989
Photograph: Dennis Smallwood

College of Art, Pratt Institute and Haystack and Penland Schools. Favouring the
apprenticeship system, he has trained many potters in his own studio. His work is
held by many museums and 'fine kitchens internationally'.

The possibility when making domestic ware of combining beauty and utility is of
paramount importance and the driving force for many a potter. Utilitarian function
and aesthetic expression need not be separated; in nature, beauty and function go
hand in hand. In the making of pottery, it is a challenge met successfully by Suzy and
Nigel Atkins.

Suzy and Nigel Atkins established their home and workshop in 1977 at Le Don,
near Montsalvy in the French Auvergne. Suzy Atkins was born in New York, USA,
but trained in ceramics at Harrow College of Art and the Sir John Cass Art College
in London. Nigel Atkins studied at the Royal College of Art in London. It was at
Harrow that Suzy Atkins first experimented with salt-glazed pottery.

She writes: 'When I was at Harrow there was interest in salt-glaze and my best
friends were deeply involved. I was very close with Annie Shattuck so I probably
caught the salt bug from her. Renton Murray and Jane Hamlyn were doing good
work too and I loved the way their pots mirrored their energy-laden personalities.

(*Above*) Byron Temple, basket, salt-glazed
porcelain, 22 cm/h, 1986
Photograph: Dennis Smallwood

(*Left*) Byron Temple, covered jar, salt-
glazed stoneware, 18 cm/h
Photograph: Dennis Smallwood

Though I have always admired certain reduction-glazed stoneware, I have never felt
it was really for me. I did a good deal of it at Harrow but I never really cared for the
"itzy-bitzy" nature of glaze calculations nor the business of breaking up the creative
flow with a biscuit firing. On the other hand I didn't want to limit myself to raw glazing in reduction. I don't think it is possible to explain objectively why I chose salt-
glazed stoneware as my creative medium. It is so much a question of personal affinity,
of the hazards of encounter and of accidents along the way'.

Since 1977 the Atkins have exhibited widely, undertaken some teaching and
been the subject of several magazine articles and craft books. In an article published
in *La Revue de la Ceramique et du Verre* of March-April 1986, Jacques Wolgensinger
wrote of the Atkins salt-glaze as being 'warm and living, of an absolute richness ...
with utilitarian function and aesthetic expression united. Why can an object not be
artistic because one can use it?' he asks.

'Suzy Atkins likes that which is simple and natural, her sculptural imagination
goes spontaneously towards useful forms. Her creative inspiration stems from daily
needs, from the essential things of life.' Wolgensinger comments that the Atkins

(*Above*) Sue Atkins, oval pitcher, gold and platinum lustres over salt-glaze, 33 cm/h, 1988

'discovered a secret that can be understood by everyone: that ceramics is an art of balance and synthesis; that the interaction of the slips, the forms and the clay is what matters most. This way, working empirically, Suzy Atkins began to elaborate her personal ceramic language. First, by a search for forms which would exalt the irregularity of the firing, and then by research into textures and surfaces that the salt would bring to life. The applied decoration is totally integrated into the plastic mass of the piece and to its place in space. The salt amplifies the reactions of earth and fire resulting in an amazing and rich association of that which is natural and that which is splendour'.

Using clay from La Borne, France, the pots made by Suzy Atkins are decorated and placed raw in their 3 m³ gas kiln, firing to 1320°C. Their basic decorating slips are mixtures of kaolins and ball clays; slips for trailing are made from 10 nepheline syenite, 90 porcelain clay, plus 10% of commercial colouring stains, pink, yellow, orange, Victoria green and mazarine blue. The firing cycle takes 30 hours and as soon as the top temperature is reached, the kiln is fast cooled to 1000°C.

Susy Atkins comments: 'Salt gives such lively nuanced surfaces and the simple slips I've developed give me a sound palette of colours and textures to build up my decoration. Of course there are disadvantages too. Fine surface quality is no certain-

(*Above*) Sue Atkins, oval wine cooler, salt-glaze, 32.5 cm/w, 1989

ty and the process is expensive in kiln ware and labour costs. So what! If you've married the right man you're not going to head for the divorce courts because he plays you up from time to time. That is me and salt, we're kind of married. We've been together fifteen years now. Fifteen good years with the same highs and lows that go with any long-haul relationship. We've solemnised our bond with more than enough affectionate oaths and although you can never be sure of anything, I reckon we're going to stay together to the end'.

On the other side of the world, and from another point of view, is the salt-glazing practice of Arthur and Carol Rosser. Their heavily salted pieces have a *wetness* that reflects their rain-forest environment, where platypus play in shady streams over-hung with green vegetation. 'Our approach to salt-glazing has been determined by our geographical situation,' says Arthur Rosser. 'We have done all our potting in tropical Queensland, remote from Australia's main population centres and virtually out of communication with other salt-glaze potters. It was natural for us to think of using local materials and we have been digging and mixing our own clay bodies for 25 years. Soon after starting to salt-glaze, using a dense firebrick kiln, we began firing with wood in the early stages of the firings, then changing to oil. Later, the challenge and pleasure of firing entirely with wood proved irresistible.

'The interaction between wood-ash effects and salt-glaze is certainly significant in our pots but we regard it as a mixed blessing, sometimes providing extra depth and subtlety to the finished pots, sometimes detracting from the pure salt-glaze effects. Our main salt kiln holds 1.7 m³ (60 cu.ft) of pots and we cannot afford to indulge in experiments on this scale, so we experiment on a few pots with a variety of slips and

(Right) Carol Rosser, lidded jar, combed
decoration, salt-glaze over high silica slips

applied glazes and try to standardise the firing cycle. Despite this, the overall results
vary more than we would like and good results come and go in cycles without us
becoming fully aware of which of the variables inevitably associated with wood-firing are most significant. Some things, such as the particular type of wood, or the
presence or absence of bark, cause gross changes, but the causes of many subtle
effects remain a mystery. In an attempt to gain a deeper understanding of the salt-
glazing process under a wider range of conditions we have recently begun a series of
test firings in a kiln fired with waste oil. The test kiln has quickly produced results
not yet seen in the wood kiln but it remains to be seen if the results can be transferred
to the wood kiln.

'Our initial excursions into salt-glazing came from a desire to make the sort of
large storage jars which used to be fired in huge wood-fired kilns in Japan. At that

time we used conventional glazes in a small gas kiln and our ability to throw large pots exceeded our skill in glazing them. Long, wood firings with natural ash effects did not occur to us as a possibility (although we have recently gained some experience in this direction) and salt-glazing seemed an attractive process. Since we are primarily makers of useful pots it did not take long for the attractions of salt-glazed domestic ware to become just as important as the firing of larger pots. It is the varied surface texture and especially the break where a lighter coloured glaze results on edges or sharp ridges that hold much of the appeal for us. These effects are particularly rewarding to touch or to see at close range as happens with tableware.

'Our natural tendency with salt is to go to extremes. We use a generous amount and temperatures up to cone 13 and obtain coarse orange-peel textures on exposed pieces. We are well aware that many subtle effects depend on using lesser amounts of salt and periodically resolve to salt lightly to allow wood-firing effects to be more dominant. However we cannot escape the feeling that very light salting is for people who do not really like salt-glaze and are reluctant to commit themselves to techniques where the salt is central rather than peripheral to the firing.'

The Rossers make the interesting point that domestic ware is judged at close range on the table, touched by hand and lip. Viewing domestic ware in a gallery or behind glass in a showcase is a poor substitute for the enjoyment of actual use. Salt-glaze, with its surface texture and depth of shine, invites this handling and close inspection. It is in using pottery that one can see and respond to the individual potter's intention and aesthetic values, and the response to the work is as individual as that of the maker. It is possibly more difficult to make good tableware than any other form of ceramics. Living with and using pottery in close domestic proximity and under such close scrutiny invites daily appraisal.

(Above) Carol Rosser, platter, incised decoration, salt-glaze over applied celadon glaze, cone 12, 55 cm/d

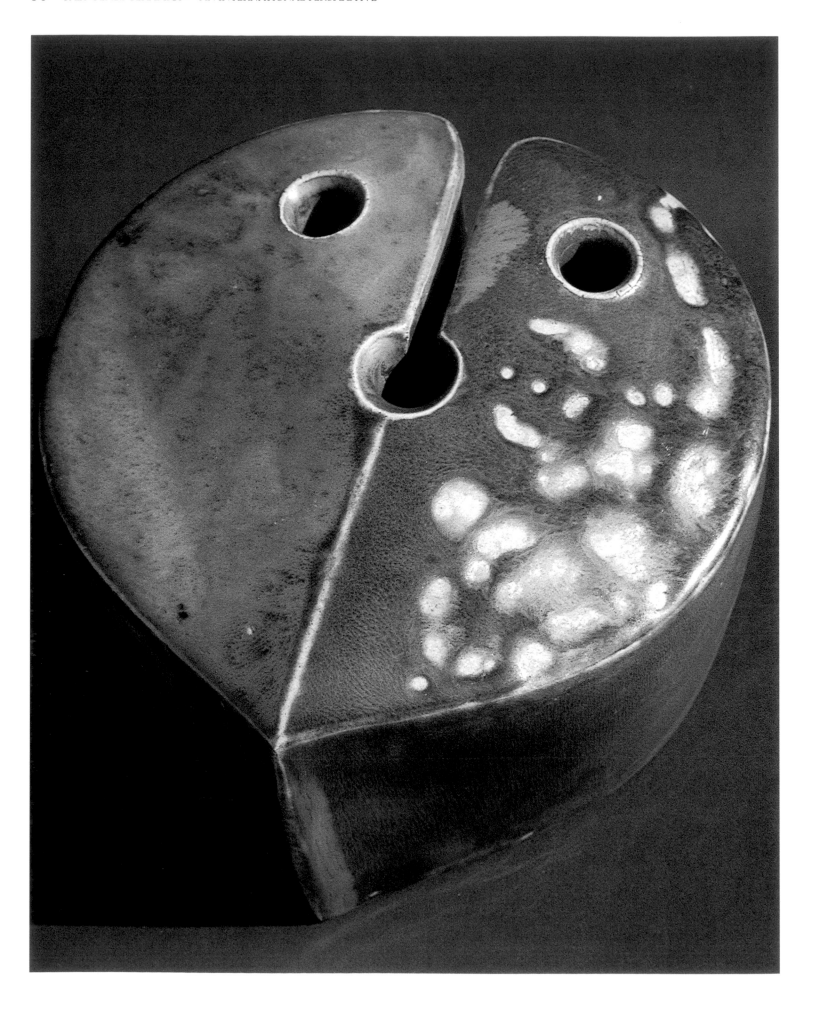

4.

THE IMPORTANCE OF TRADITION

'To transform known traditional techniques into new and exciting possibilities for the potter and to make pottery that is useful for individual demands or needs today — that is my goal in working with salt-glazed stoneware.' Wolf Matthes trained as a master potter and ceramics engineer at the Staatliche Ingenieur and Werkschule fur Keramik in Höhr Grenzhausen, Westerwald, West Germany, an area vital in the history of salt-glazed stoneware. He had an early interest in geology and minerals, and after working as a potter for some years in Switzerland he became a ceramics engineer in a Swiss brick factory group in the field of research and development. Since 1970 he has been a full-time teacher at the Höhr Grenzhausen State School of Ceramics (Ceramic Design) in technology, glazes and design but has also continued to work in his own workshop. His main interests are in research in the field of salt-glaze stoneware techniques, the development of colour in glazes and slips, and bodies for salt firing. The results of his studies have appeared in many publications on ceramics. He continues his investigations using natural materials, minerals and rocks for stoneware and into the history of salt-glaze ceramics.

'Salt-glaze is an adventure,' he says, 'always surprising, and unexpected results occur when I give the kiln and the salt some leeway to work, independent of my will and tendency to perfection. In this way, despite long years of experience, I can be curious about opening the kiln, be astonished at the result with wide-open eyes, or be endlessly disappointed. The firing in the gas kiln is simple and routine can quickly set in. There I use my experience to repeat good results and I can usually predict what will come out of the kiln; there is less opportunity now for the astonishment that happened to me as a ceramic beginner. Yet with salt-glaze, the stoneware receives life and unplanned variety, different from the routine, and work with salt-glaze becomes fun again. A wood-fired kiln would be better still for it, insofar as I don't want to make standard catalogue work'.

To safeguard the traditional values of salt-glazed stoneware and keep them from degenerating to mere decorative or souvenir items, Wolf Matthes says that young ceramists using the technique, 'must wish to awaken the old forms and decorations to new life'. In his writing and teaching, Matthes encourages others in original research. In his own work he looks for new possibilities in ceramic art while appreciating and acknowledging the important work that has gone before him. In this way, he believes that ceramists can build a tradition 'but equally, should develop new and

(Facing page) Wolf Matthes, wall piece, salt-glazed stoneware, basalt glaze, 1250°C, 28 cm/d

(Right) Gerhardt Hemberle, pouring vessel
Photograph: Ferdinand Joesten

contemporary ways of using materials and techniques in their own specific ways'.

Gerhardt Hemberle, of Habsthal in Germany, brings to his work the sympathetic understanding of the traditional German salt-glaze, looking upon his team of workers in the pottery — himself, his wife, Ulrike, and four journeymen/apprentices — 'as handcraftsmen potters, living and working under the same roof'. In 1985 they established their pottery in a 16th-century mill situated by a brook, with old trees and space enough to breathe and work.

'It is our goal,' he writes, 'to maintain the simple and unpretentious handcrafted practical forms as they have been brought down to us over the centuries. To discover them anew and influence them anew will, I hope, be our contribution. In doing this, we can link the traditional techniques of salt-glaze stoneware of Germany and France. The type of production of our pots is as important to us as the pots themselves. The pots and the vessels should be useful and fulfill their function here today. Comparable to the importance of function is the importance of aesthetics. We prefer simple forms and line designs, using simple tools to beat and cut the clay, a *Malhorn*

(slip trailer) and brushes for decoration, and our own ash glazes and engobes. We can
direct the atmosphere in our self-built kiln with the greatest precision, and our technique of salting gives us additional means of achieving the glaze and surface formation that we are seeking'.

(Above) Wolf Matthes, jar, salt-glazed
stoneware, celadon glaze, 1230°C, 22 cm/d

The traditional techniques of Germany and France continue to be relevant to
potters interested in salt-glaze, forming a source of inspiration. Although times have
changed, the spirit of these old craftsmen lives on. The old potters of La Borne in
France learned their skill at a young age from their fathers and grandfathers. At the
height of the La Borne pottery activity the population was 800, the roads were alive

(Right) Elisabeth Joulia, platter

with horses and carts carrying wood and clay and almost every day a kiln would be firing. The La Borne style was easily recognisable. After World War II there was no longer a demand for their style of utilitarian wares or bricks because of the availability of factory-produced goods. However there was a market for decorative pottery and, while some potteries disappeared, others remained and new potters arrived to meet the demand. Today the population is approximately two hundred.

The new potters worked among the old craftsmen and learned from them, but they had a different view of pottery. The accent of their production was on the satisfaction of the creator and the pleasure of the possessor. These artist-potters gave exhibitions and attracted critics and collectors to La Borne. They were pioneers who made the bridge between the old La Borne and the La Borne of today.

During the 1960s the name of La Borne in the world of ceramics was re-established and it became a mecca for potters coming from all over the world. They came to visit or work for a while and many stayed and settled down. La Borne now has an international population of potters and, slowly, old deserted workshops are being used again. New kilns of all different types are built: kilns from Sevres, Japanese three-chambered climbing kilns, Korean kilns, kilns varying from 1 m³ to 12 m³; individual kilns for individual potters, all having their own style and character.

One potter who came to live in La Borne was Eric Astoul. He built a wood-firing kiln, similar to a traditional kiln from Thailand, with a capacity of 7 m³. Using an

(*Above*) Jean-Claude de Croussaz,
container, salt-glazed, wood-fired

(*Left*) Elisabeth Joulia, teapot

iron-bearing clay, Eric Astoul's wares rely on the ash from the fire and the flashings of
salt vapour for decoration. A potter who visits La Borne regularly is Jean-Claude de
Croussaz from Geneva; every summer he spends some time making works and firing
them in the salt-glaze kilns. De Croussaz teaches ceramics at the Ecole des Artes
Decoratif in Geneva, and is well known for his highly decorated exhibition pieces.

Elisabeth Joulia has lived in La Borne since 1949 and saw the rebirth of the ceramic activities after the Second World War. Three years later she organised her own
workshop where she still works and from where she has had considerable influence
on other artists.

For an exhibition in 1989 in Emmendingen-Mundingen, France Kermer wrote of
Joulia and her work: 'She embodies a many-sided artistic personality which has
made, and continues to make, its mark on the ceramics of France. Her repertoire of
forms extends from vessels to the free-form sculptures whose organic lively forms she
shapes in a continuing growth process like a being of nature. She experiments with
various techniques and searches for her clays nearby or from distant areas. This
method of searching for particular clays is important to her for the earthy qualities
and the uniqueness that she wishes to bring to her work. "A ceramic work can be
read when the clay has been treated as a human being rather than a lifeless carrier
hidden under a thick glaze." These words of Elisabeth Joulia yield a meaningful view
into her work. Her creative activity is based on the pure unembellished material, a
traditional and aesthetic value of the "school of La Borne" '.

Joulia continues, as Kermer wrote, 'to listen to the material, feeling, observing
with sensibility and sensitivity for the nature of things ... her very feminine work
radiates a great power. In 1983, the "grande dame" of French ceramics as she is today,

(Right) Elisabeth Joulia, *form*, white
stoneware, wood-fired, 1300°C

was honoured by a retrospective at Musée Municipale St Amandles-Eaux; her international presence contributes to the importance of French ceramics beyond the bor-
ders of her country'.

Christine Pedley first heard about La Borne when she was still studying at Harrow
School of Art. At Harrow she followed an intensive, practically orientated pottery
course from 1964 to 1966. She and Janet Stedman were both fascinated by the
stories they heard about La Borne and its huge wood kilns and they knew they wanted to go there and look for themselves. While Stedman went directly to La Borne,
Pedley first studied with Jean-Claude de Croussaz in Geneva and Antoine de Vinck
in Brussels. Stedman wrote enthusiastic letters about La Borne, the wood firings, the
people, the possibilities and the inspiration of the place, and in 1968 Pedley arrived
to work in the same studio.

(*Above*) Christine Pedley, platter, slip decoration, salt-glazed stoneware

She writes: 'My first experience of using salt in the kiln was with Pierre Digan and Janet Stedman with whom I worked for the first 18 months. Influenced by the traditional potters, we used small once-fired pots (the size of large egg-cups) placed throughout the kiln in the protected areas where wood ash vapours had more difficulty reaching. These salt pots gave a flash of local salting on pots which would have otherwise remained matt and dry on an unglazed body. When setting up my own pottery in 1970, I continued to use the salt pots in the same way in my small downdraught wood-fired kiln. I would also throw in a handful of salt behind each bagwall at the end of the firing for an extra sparkle. The total amount of salt used would barely come to 500 gm for a capacity of 800 L of pots. These firings I referred to as "semi-salted" wood firings.

'The first full-salt firing I did was in Japan in 1974 while working with a traditional pottery family in Mashiko. I travelled overland through Asia with my husband, Steen Kepp, and shortly after arriving in Japan we were accepted as guest potters and asked if we would fulfill the project of building a wood-fired kiln for salt-glazing, producing European pots to fill it. We had two successful salt firings. Returning to La

(*Above*) Shimaoka Tatsuzo

Borne we were inspired by this experience in Japan and immediately did a full-salt firing in our wood-burning kiln. After reaching 1300°C we allowed the temperature to drop to approximately 1100°C, then we closed the chimney and filled the firebox with wood and let the whole kiln choke (the blue-brick or downhill reduction technique). The pots came out silver-grey with mauve tints in some places. They had a rather cold metallic look about them but we found them quite special.

'Later I started using a fibre kiln, gas fired, as I wished to stay as close as possible to the naturalness of the unglazed, local La Borne clay body in contrast to the preciousness of the ash-glazed pieces. Slowly the amount of salt crept up, the use of slips became more varied and the ash glaze was limited to the interiors of the pots. I now use 3 kg of salt introduced beside the burner block, blown in slowly with the aid of a vacuum cleaner. The salt is evenly distributed through the kiln eliminating the need for salt pots. Salting starts at 1100°C and lasts about one hour. I fire up to 1290°C, the whole firing cycle lasting about 8 hours.

'Salt has become an important part of my firing technique and, as with wood, it brings out the best in the clay, giving warmth and colour. My pots are basically functional, except for the large decorated dishes which I like to see hanging on the wall as a painting. There is a feeling of fulfilment when I succeed with one of these large dishes. I studied drawing and watercolour painting before I discovered clay and when the potter and painter in me come together I am happy. I use slips, and with large brushes, fingers and combs work on the flat surface before the plate is turned. The decoration seems to belong; these are the ones that feel right, but they do not happen often enough.'

Christine Pedley is active in the community of potters in La Borne and though working alone, she allows students to come to her studio. She plans to stay in La Borne with her two children, making a living from her salt-glaze ceramics, and helping to promote the atmosphere of the village with its international pottery population. Visitors savour the history of the village, inspect the museum with its fine old pieces and enjoy the new work that is being made and the pottery activities which the village organises. Writing on La Borne, Murray Fieldhouse, editor of *Real Pottery* magazine said: 'It is a region from which one rarely departs, indeed, whoever comes there for a day runs the risk of spending the rest of his days there; he is adapted and adapts...'

Two traditional potteries in Sweden have been taken over by enthusiastic young potters looking to rekindle interest in the early salt-glazed wares of their country. The Walkara Stoneware factory 'Handmade from Clay and Salt since 1864', south of Heisingborg in the valley of the River Råån, is operated by Asa Orrmell, Ulf Jonsson and Bente Brosbol Hansen. 'A visit here,' they say, 'is like stepping back 125 years into the past'. They dig their own clay on site, wash and filter it, and prepare it for throwing. The pots are stacked in the kiln, separated by small grooved clay pads to stop them from sticking together. When the kiln is full, it is fired for three days, to 1300°C. After one week, the kiln can be opened to reveal the classical pots that are traditional to it, pots used for storing, as they say, 'everything from jam to schnapps'. Creating a nostalgic atmosphere, these potters are working hard to make vessels in the time-honoured way.

Not far away at Raus, Lars Andersson is creating a similar workshop from an old factory. Originally owned by two potters, the last surviving one still comes to the factory nearly every day to watch the progress and discuss old times. A former school teacher, Lars Andersson says the old pottery has a good feeling, with an energy and a life. It takes him four months work to fill one of the two kilns in the pottery, both of which are in good order. Taking two weeks to stack the kiln and then 70 hours to fire

(Above) Shimaoka Tatsuzo, plate, impressed decoration, 58 cm/d

it, he uses 7 tonnes of coal, salting twice, throwing two or more shovel loads of salt into each of the seven fireboxes around the perimeter of the kiln.

Idealism and a sense of continuing tradition are giving these Swedish potters worthwhile goals, enabling them to create a life style which, although often hard, is fulfilling and also satisfies their need to earn an income.

In Japan, the tradition of salt-glaze is relatively recent, and it is due to the influence of Shoji Hamada, who became one of Japan's 'Living National Treasures' for his work as a potter and for his promotion of the *Mingei* (art of the people) movement, who inspired others to see the beauty of salt-glaze. Hamada had brought back some

Bellamine-style salt-glaze jars from England when he returned to Japan after working with Bernard Leach in the 1920s; but it wasn't until after his trip to America in 1952 that he began to use the technique.

He has been reported by Bernard Leach in his book *Hamada* (Thames and Hudson, London, 1976) to have said: 'At the beginning all my salt pots appeared rather western, to my surprise. But after many years, I feel my salt-glazed pots have mostly become Japanese. The method is originally German, the clay is Mashiko, and the salt is an imported coarse salt from Spain. There are many hazards in doing salt-glaze, for salt will attack the bricks of the kiln itself, causing material from the ceiling of the kiln to drop down on the pots. Salt sometimes penetrates the body and causes bottles to bend and stick to one another. However, it is because the risks are great that I have found great stimulation in using this technique. After sixty years of making pottery one can become too comfortable and proficient. It is the very challenge of this direct, primitive method, with all its possibilities of failure, that has proved an exhilarating experience for me'.

Shimaoka Tatsuzo, who studied at Hamada Shoji's workshop for three years, from 1946 to 1949 , is today one of Japan's most noted potters. In 1953 he established his own workshop and kilns at Mashiko. Like Hamada, Shimaoka is a supporter of the *Mingei* movement in Japan, a movement that espouses a definite philosophy, as outlined by Sir Hugh Cortazzi in an article on Shimaoka (*Arts of Asia*, March-April 1985). '*Mingei*,' he says, 'can be summed up in the belief that pots are made by a constant repetition of the movements of the body. It is this movement, rather than any conscious effort, which produces pots in which the potter was not conscious of himself as an artist; rather, his pots were made naturally as a part of his daily life'.

In *Mingei*, the potter does not think about his image, nor of commercial values; he makes pots because of the feeling within him rather than them being the product of intellectual endeavour. He makes pottery for use within a tradition of potters, naturally producing objects of beauty. Shimaoka, although steeped in the tradition of *Mingei*, has combined tradition with his own personal dimension in the way of a true artist and has developed a distinct individual style. One of his special techniques is the use of an impressed rope texture. A coating of slip is painted over the impressions left by the rope; when the slip has partially dried, the pot is scraped to reveal the inlaid decoration. A variety of effects can be achieved in this way.

Salt-glaze is a speciality of Shimaoka. He says: 'I like the bright colours I can get from salt-glazing. As with my other work I use rope impressions and inlay a cobalt blue slip or a high alumina white slip in which I put ordinary kitchen salt. When fired in a salt atmosphere, this white slip turns into a rich warm brown. I fire the last chamber of my four-chamber climbing kiln, throwing in rock salt when cone 9 has dropped. My cone 10 wood-fired reduction pieces can be dark, even sombre, but I like the bright feeling of the work that wood ash and salt combine to make'.

Shimaoka has exhibited in many countries and his work is familiar to collectors of ceramics throughout the world. In it they can see a true balancing of tradition, drawing on work from the past, using his own creative ability to produce works of contemporary beauty.

Absorbing and using tradition as a source for innovation, and building on it anew, has been the inspiration for many artists. The responsibility for today's potters is in taking a role in that continuum: making, adapting, establishing one's own style and becoming part of a future tradition. The ceramists discussed in this chapter are aware of the importance of tradition; they recognise the powerful influence it has had on them and that their own work will continue to touch and inspire the work of others.

(*Above*) Shimaoka Tatsuzo, jar, impressed
decoration, inlaid slip, 37 cm/h

5.

CONTROL OVER UNPREDICTABILITY

As many of its practitioners will tell you, one of the characteristics of the salt-glaze process is its unpredictability. For some, this is part of the challenge and excitement, not wanting complete control, but an element of mystery and surprise. For some ceramists, however, control is sought and, after considerable experience, achieved. For many of these ceramists, the style of their work is also under control, usually refined, even hard-edged, with strict lines and precise patterning; and this work is often in a muted colour range. Yet for them, the salt-glazed surface is the preferred one to complete the total aesthetic concept. This chapter deals with a number of ceramists who seek this control over the firing process.

Walter Keeler has been called a sophisticated potter. Influenced from an early age by Roman pottery and glass, his ceramics, by their forms, reflect his English background. He trained at Harrow School of Art, establishing his first workshop in 1965. His present studio, near Penallt, Monmouth, in south-east Wales, was established in 1976 in partnership with his wife, Madeleine, who is also a potter. With many exhibitions and awards to his credit, workshops, and articles published about his work, he is an artist with an international reputation.

He says about his work: 'Throwing is my primary technique, and the inspiration for my pots. They all begin on the wheel, though most are altered in some way, or assembled from various components. Pots released from their base no longer remain circular. Rims are cut away to leave raised pouring lips, or a vessel is set back on its base, as if affronted by something in its path. Handles may be pulled and sensuous, growing from a rim, or extruded with metallic accuracy, appearing just to rest on the pot. Spouts, whether thrown and altered or pressed in a mould, contribute their peculiarities to the assembled pot. The goal in this complete process is the finished pot performing its function; a surprising object doing a commonplace job'.

Walter Keeler uses the wheel and kiln and other tools with confidence and sympathy. In an article for *Ceramic Review 77* in 1982, Michael Casson described Keeler as 'deft' with 'methods carried out with masterly skill. Many of his influences come from hard-edged objects. It is a measure of his ability that he can walk that tightrope of articulated forms and assemblage methods and yet keep the pots dynamic. He keeps his balance because of a combination of extremely sensitive skill, the relaxed humour of the forms, and a deep-seated understanding for the enduring qualities of clay vessels, their forms, their right weights, their colours and surface qualities ... Keeler sees himself as a traditional mainstream potter concerned with

(Facing page) Volker Ellwanger, jar

(*Above*) Walter Keeler, macaroni teapot, 16 cm/h, 1989

classical forms and constant values.'

Keeler has also been described as a 'modern potter'. David Briers, writing for the *Ceramic Series* of the Aberystwyth Arts Centre, said Keeler 'now possesses an enviable technical expertise' and 'tends his kiln with the experienced intuition of an improvising musician'. In another article, for *Ceramic Review 112*, Briers described Keeler working with great care and precision, emphasising the attention paid to rims and edges, the unity of the salt-glaze texture throughout his range of designs, which 'once established are only altered or adapted quite slowly'.

Keeler added some technical information to the article: 'All my pots are functional, it is a fundamental justification and a challenging starting point. If the pots could not be used I would not bother to make them. My forms are derived from working with the clay and the processes of throwing, turning, pulling, extruding, pressmoulding etc. I do not deny visual links with tinware (but these are not contrived) or early industrial potters and peasant wares. I believe in a certain integrity of making, which derives from the time I made production wares. I also believe firmly that tradition holds a key to innovation. I hope to echo the qualities in the old pots that delight me, to emulate the inventiveness and sense of humour of 18th-century industrial potters rather than attempt a kind of reproduction'.

Keeler's biscuit pots are glazed and slipped, with the slip coating giving a smoother (less orange-peel) finish after firing. The exposed body yields a definite orange-peeled texture. Colour, using mixtures of oxides and stains, is sprayed on both surfaces, most commonly consisting of black stain, black stain plus cobalt, or cobalt and chrome oxide.

(Above) Walter Keeler, teapot, 16 cm/h, 1989

The artist has two salt kilns, the larger one 0.9 m³ (32 cu.ft) setting space and used most at present, and the smaller one about 0.2 m³ (8 cu.ft) which enables a more frequent firing interval. They are both catenary arch hard brick structures, with 5 cm of ceramic fibre for outer insulation. Each kiln has two oil burners in opposite corners; the big one uses swirlamisers and the small one has homemade burners running on compressed air. The firing cycle in the small kiln goes from 0°–600°C in as little as 15 minutes. At 1000°C (Orton cone 06) reduction takes place. After five and a half hours, at 1250°C (Orton cone 9) salting begins using about 6.5 kg (15 lb) salt. This usually takes about an hour and a half, and then the kiln is re-oxidised. The temperature drops during salting but is brought back up to 1280°C (Orton cone 10) which takes half an hour. The total firing time is 7 or 8 hours.

Keeler's clay body is a mixture of 75 SMD ball clay, 25 Moira stoneware plus 10% sand. Pots are biscuit fired at 980°–1000°C. He uses an engobe (biscuit slip) consisting of 60 felspar, 40 china clay. An interior glaze (Phil Wood recipe) is 70 Cornish stone, 30 Wollastonite, 8 red iron oxide.

Looking at a Walter Keeler teapot with its precise lines, articulated changes of direction, and overall sharpness and presence, one feels the ceramist has the materials, processes and ideas firmly under control. Yet the softness of the salt-glaze surface, the challenge of technique sought by the potter and, often, a small quirkiness or perkiness to the form that gives the work life, endows a friendliness to a rather reserved piece.

Aage Birck, from Denmark, is a potter who has 'travelled through clay from low-fired earthenware to high-fired stoneware, from functional pots to sculptural works, from small electric kilns to huge self-built gas kilns with a sidestep to *raku*'. He likes to know that the techniques he is using will suit the concepts he has for his work. He says: 'I aim for a maximum control of the firings, and pieces which miss that control aren't left many chances of passing the needle's eye. The *raku* technique was fascinating because of the dramatic changes of the material from different firings combined with different ways of cooling, but with my background in the functional pot, I never became familiar with the poor density and strength of the fired clay and missed the ringing sound and cool touch of well-sintered clay.

'At that time my attention was drawn to salt-glaze, walking through museum collections or "running" through magazines. As the hungry bird sees the worm on the ground or the man a restaurant in the city centre, my attention was drawn to German traditional salt-glazed stoneware in brown and grey colours, and the French in all the shades between, depending on the clay, firing methods and ash deposits. I became excited by the orange-peel texture, the flame flashes and the underlining of finger marks, decoration and edges.

'We, that means Heidi Guthmann Birck and I, who have been sharing clay, bread and breakfast for 25 years, built a small test kiln for salt firings in the garden, bricked up without mortar and with steel wires tied hopefully around the whole structure to keep it from falling apart. And it worked: one could see that salt fumes had passed through the kiln touching the test pots and tiles. We were encouraged enough to build a permanent salt kiln under a roof, and we continued making salt-glazed stoneware and experimenting with it. Description of these experiments, technical information and some results are featured in our article "A Way of Salt-glazing" published in *Ceramic Review 82* in 1983.

'Since that time my aesthetic relations with salt-glaze have changed. Being a novice in a new medium, one can be taken in just by what comes from working with it, especially with clay. Few other materials turn your fingerprints into a sculpture by just squeezing a lump of plastic clay, few other colours make such dramatic interactions as layers of different glazes fired to their runny stage, and few materials can be transformed as radically as the firing of clay, especially with pit-firing and *raku*-reduction, and also wood and salt firings. I think that all this may have an overwhelming influence, even on trained artists and craftsmen which, in the beginning, confuses their ability to judge the result. That is maybe the explanation why an amateur's work in clay and the clay work from artists coming from another field than ceramics have a similarity. It takes time to tame the "dragon", to weed away empty effects, to figure out which of the many coincidences, all part of salt-glazing, give an advantage or a disadvantage to the fired piece. It takes time to fit the process to one's artistic will, if, at least, one believes that intellect is a part of art.

'After a period of experimenting, trying all the things I could imagine and read about, I aimed for the orange-peel texture of heavy salting, occasionally combined with partial ash glazing by dusting exotic ashes where I liked them to be. The colours of titanium yellow, reduced into blue and black on the flame exposed parts of the pots and the shades of dark-red to bright-orange from aluminia-high slips especially attracted me. The forms were simple with few additions and decorations. I changed the forms slowly, towards greater complexity, often uniting several forms combined with wood, and the effect of heavy salting became unsuitable. The sparkling surface diverted the attention from the form, the face disappeared under too much make-up. This new work needed only a touch of salt, like the sea wetting the stones at the shore developing the colours of clay and slips.'

(*Above*) Aage Birck, container, stoneware and wood, slips, wax resist, sprayed with salt, 1300°C, 40 cm/h

When the Bircks moved from the open countryside to the centre of an old town, salting was impossible for almost a year, until a new salt kiln was built with a colleague. In that period they started, and still continue, to evolve stoneware slip glazes with a much dryer surface 'like unglazed earthenware, unevenly reduced, marked by years of use and forgotten in the soil, with the muted colours of grey, brown, beige, and red'. All the tests were partially sprayed with a salt solution and some gave interesting and surprising results. 'Boring white slips changed to lively grey-orange to shining charcoal-black, others went from yellow over greenish to dark grey depending on the thickness of the salt layer. Also, the texture was affected. Sometimes

(Right) Aage Birck

the slip in certain areas was eaten away and the body made visible and flashed by the salt.'

Aage Birck queries whether this is salt-glazing or not. In some aspects it is. The shiny areas, where the salt layer is thick, is evidently a glaze. The influence from the salt on the colour is similar to that which can be seen in lowfire-saltings, more a flashing than a glaze. He comments: 'Without success I have tried to get similar results by spraying layers of high alkaline glazes but this gives no sign of orange-peel or the underlining of rims and marks. This method of applied salt fits to the way I want my work to be at the moment. This way of applying the glaze gives possibilities for using different resists, and my influence and control on colour evolution and placement is higher than in a salt fire. But the drama is less'.

Drama is an inherent part of the traditional salt-glazing process: physically tossing the salt into a hot firebox, hearing the crackle as the salt spits and volatilises and seeing the vapours rise into the air. However, the effects of using salt for changing the surface of clay or engobes in a controlled situation is now the favoured technique used by a number of potters. André Bertholet from Switzerland and Ernst Pleuger, from Germany, winner of the 1989 Salzbrand exhibition in Koblenz, both choose to place their pots in saggars inside a regular stoneware kiln. They seek the combustion of the reduction materials, charcoal, sawdust etc. plus salt to flash their pots. The resulting vapours and carbon inclusion alter the colours of the clay body or applied slips to form bands of matt, unglazed blacks, greys and reds. The use of copper oxide and ochres in the slips can intensify these effects.

Using the gloss of salt-glaze over already applied glazes is another way to exert control. Volker and Veronika Ellwanger, ceramists from Lenzkirch in West

(*Left*) Aage Birck, container, salt-glazed, wood-fired stoneware, 40 cm/h

Germany, use the techniques of salt-glaze on applied glazed pieces. In this way they achieve a range of colours not obtainable with the salt alone. They share a kiln and both use similar glazes applied to their pieces before the firing. However, the contrast in their individual works, in the intention, in the type of production and in the essence of the shapes they make, is marked.

Writer Franz Disch, in describing their work says: 'Naturally one arrives at similar results by the use of certain materials and by their application at commensurate temperatures and burning conditions. Salt-glaze is one of the many possibilities which both Ellwangers use. Rarely do they use salt-glaze alone and one finds in Volker Ellwanger's ceramics exemplary examples of *celadon*, *clair-de-lune* and *sang-de-bœuf* glaze types. Independent of the Asian model, he has also developed his own glazes and through experience has found an individual interpretation corresponding to our own time through form. In the interest of the continuation and development already made in glaze technology, and because he is a teacher, he strives for a pedagogical message; he does not say to his students "it is done this way", but is more concerned to encourage his students to discipline themselves. These self-set bound-

(*Above*) André Bertholet, vase, saggar fired, 1989

aries do not mean a restriction for him but rather give him the freedom to find formal variations'.

Professor of Ceramics in the Faculty of Visual Arts in Mainz University of Johannes-Gutenberg since 1984, Volker Ellwanger trained at a pottery and tile oven factory, and at art schools in Darmstadt and Berlin where he received his master's diploma. He set up his own studio in 1961 where he was joined by Veronika Ellwanger in 1967.

Veronika Ellwanger assembles her vessels from discs made from grogged stoneware clay. The resultant piece is conceptual and the forms are architectural, reduced to squares and cubes, the openings resembling an entrance to a cave or a door to a house. Franz Disch comments: 'While their approach is disciplined within the heritage of salt-glaze both develop their own expression or interpretation. Thus, they bring a new dimension to the traditional salt-glazing which has been known in Germany for 500 years'.

It is this search for control over the unpredictable process of salt-glazing that motivates some potters to challenge the medium. This is the case of Dutch-born potter, Johanna DeMaine, who now lives and works in Queensland, Australia. She writes: 'I am aware there is chance and inevitability in the salt-glaze process but systematic control dictates my approach to salting. Everything is predetermined with nothing left to chance except the flame itself. It is this sense of control which helped determine my interest and direction in this technique'.

'My heritage, together with travel and study, have resulted in many influences on my work. I grew up with magnificent German Bellamine jars and blue and grey salt-glazed pieces that my parents had acquired whilst living in Holland. Travel in the East has given me an appreciation of simplicity of form and decoration and a combination of these eastern and western influences have surfaced in my work. I strive for forms that are simple but elegant. I have a great affinity with the classic forms of the Chinese.'

Writing for *CraftArts* magazine in 1987, Jeff Shaw recognised this characteristic of Johanna DeMaine's work: 'Despite local and influential contacts, it would appear that the theme of the DeMaine work relates to an older European salt tradition because of the robust simplicity evident in her work. Leach, in describing these qualities, spoke of the simple dignity of the old tradition, clean and sober in form and clay-like in throwing. These qualities are essentially to be found in the DeMaine work, with strong, simple forms frequently enlivened by vigorously sculptural faceting or fluting, enhanced by gently salted surfaces with subtle colour variations'.

Johanna DeMaine says of her work: 'The pots are either fluted or decorated with slip trailing. I have developed considerable control with these two techniques and now am exploring and enjoying the endless permutations which can be achieved with them. Sometimes I use a slip made from the red mud which is the waste by-product of bauxite refining. It is a heavy iron mix and gives a rich burgundy colour to the pot. Also, it "orange peels" beautifully. I have also used glazes in the past to highlight areas. As I raw-fire all my salt work, I use suitable glazes. A *tenmoku* will leach out to a golden yellow, while a magnesia glaze turns to a jade-like *celadon* under the right conditions. The pots are made from either an iron bearing body or a white stoneware. After a painful lesson in cystabolite dunting, I mix various clays to produce a more open clay body to overcome this problem.

'My kiln is top loading, has a capacity of .3 m³ (12 cu.ft) and is gas fired, using two 5 cm venturi burners which direct the flame under the floor before hitting the back wall and then travelling through the kiln in a classic downdraught pattern. I used dense firebricks that had come from the calcination kilns at the alumina plant in

(Left) Veronika Ellwanger, vase, salt-glaze over applied glaze

Gladstone. These bricks are composed of 75% alumina. As the salt is dropped through the top of the kiln directly into the flame path, I used a salt resistant castable as a false floor to prevent corrosion of the floor bricks. The bricks were in such good condition after more than 60 firings that the kiln was demolished and the bricks reused when we moved to Landsborough. Two 30 x 45 cm silicon carbide shelves side by side are used in the kiln in layers. I use a mixture of 80 alumina, 15 kaolin, 5 bentonite as a wash on one side of the shelves. I don't wash the underside as the wash tends to flake off and fall on the pots below. I clean the bottom and sides of the shelves with a wire brush after each firing as a froth builds up and this could also drip on to the pots underneath. The shelf props are dipped into the wash to prevent them sticking to the shelves. I also use a wadding mix of 50 high alumina fireclay and 50 fine silica sand.

'My work follows two directions, both related to the salting process. One firing I will salt, the next firing I rely on the residual salt in the kiln to *blush* the pots. Thus I achieve two different types of finish, both equally demanding to obtain the result. Both of these rely heavily on the pack and the consistency of the firing cycle. The

(*Above*) Johanna DeMaine, jar, salt-glazed stoneware, fluted decoration

pots are either set on shells or wads. For the blush firing I stack the pots separated by wads. The stacking and placement determine the flame paths and this is indelibly marked on the pot, the white or lighter areas having been masked or protected by another pot. The black areas represent carbon trapped in the body.

'I have a 30-hour firing cycle for the quality of surface I seek. The kiln is soaked for 3 hours at 900°C to burn out all the carbonaceous materials before reduction starts. This avoids the problem of bloating. An hour of extremely heavy reduction follows, then a medium to light reduction to top temperature takes a further 10–12 hours. When cone 11 starts to fall, I commence salting. I leave the kiln in medium reduction during salting and use 2.5 to 3.5 kg salt in 0.5 kg baitings every 10–15 minutes. After the salting is finished the kiln is re-oxidised for half an hour to clear all the gases and then the gas is shut off. The kiln is allowed to cool rapidly to 1000°C with all the ports and the chimney open and then the kiln is clammed up tightly. It usually takes three days to cool down because of the density of the bricks. For the *blush* firing I use exactly the same schedule but I don't bait with salt. It is simply the residual salt coming from the walls which give the gradations of colour.

'Recently I have been working with the addition of copper carbonate to the salt. This produces a very soft mauve to pink on the white stoneware fading into a soft grey where the pots are shielded by others. I am now also using cobalt slips and cobalt oxide added to the salt to produce a soft blue orange-peel surface.

'I am continually excited by the possibilities that can be achieved by salt-glazing and, although I have been working with the technique for 11 years, I feel I have only just started. It seems endless and I cannot foresee that I will ever tire of the salt-glaze process as so far I seem to be in control.'

The ceramists portrayed in this chapter have considerable technical expertise. That they choose to accept the challenge of salt-firing, with its unknown quantities and surprises, is a paradox. While they require control over their medium, they also use to advantage any blessings the kiln has to offer. This makes their work an even more individual pursuit of their artistic intention.

(Above) Johanna DeMaine, jar, porcelain, residual salt, slip decoration

6.

A LOVE OF FORM AND SURFACE

The surface of salt-glaze is typically known as *orange peel*. It has also been referred to as *tiger* or *snake skin*. These names conjure up a textured surface which is quite unique in ceramics. When salt is thrown into a hot kiln, it volatilises. The chlorine component escapes through the chimney, combining with the water vapour from the combustion or in the atmosphere, while the sodium seeks out the silica particles in the clay body of the pieces being fired in the kiln. This sodium and silica mix is quite molten at high temperatures and as the ceramist throws in more salt, the sodium builds up on these sticky melted areas until they become raised from the surface of the clay body. This texture may entirely cover the piece of ceramic or it may be isolated to exposed areas. Masking areas with slips or protecting parts of the work by stacking pieces closely together can shield it from the salt fumes; these areas will develop less orange-peel texture.

It is this contrast in texture that excites many ceramists to experiment with the technique. Not all clay work is suitable for salt-glaze treatment and experimentation with form usually goes hand in hand with experiments with the surface. Sometimes it is a concept of form which leads the ceramist to salt-glaze, the wish to reveal the clay and not detract from the overall impact of the piece. For others, the surface leads the way. Ultimately, both form and surface should produce a coherent statement of the ceramist's intention.

For Bente Hansen, of Copenhagen, Denmark, salt-glaze is the element which unifies the surface and the form. In writing an introduction to an exhibition held at the Graham Gallery in New York in 1989, William Hull, Director Emeritus of the Museum of Art at the Pennsylvania State University, said: 'The integration of decor to form has always been a hallmark of Bente Hansen's work. Strong, complex geometric designs with bold colour fields enhance the architectonic quality of the pieces. Bente Hansen's mastery of form, ornament and surface, achieves an authority in her œuvre rarely attained in today's ceramic art. That her work is salt-glazed adds an intriguing dimension. When one considers the strength of her form and ornament, the use of this ancient glazing technique seems completely compatible, recalling the timeless quality of medieval German pottery'.

Bente Hansen states: 'The reason why I salt is the firing process. It gives the pot a unity; I am involved all the time from the beginning, wedging the clay, to the end of the process, opening the kiln. Strong, brilliant colours are available in salt, not covering the surface with a layer of glaze, but revealing it.

(Facing page) Bente Hansen, container form, 22 cm/h, 1988
Photograph: Finn Rosted

(*Right*) Bente Hansen
Photograph: D. Kogh

'When I started twenty years ago, making ceramics in the basement of an historic building in the very centre of Copenhagen, I built a gas kiln for salt. Except for a short period at the Royal Copenhagen factory, I have made only salt-glazed ceramics. I often think about an easier way to obtain the expression I want but a salt-glazed piece will be, for me, forever a wonder of beauty and mystery. Even when a firing is not a success, there will be nearly always, even on the most miserable work, spots where the eye can disappear into another world. When I started, it was the form first of all where I put the most effort. I made rather organic forms, thrown and altered. I worked in a movement from convex to concave waves. The decoration followed form and supported it. The salt was a thin layer, only just enough to develop the colour.

'About 6 or 7 years ago I started to decorate against the form. I began to work with the contrasts, the soft form with sharp-edged decoration. My pots became more and more simple in shape, concentrating on the proportions and the strength of the curved line. I'm creating, but not without tradition in mind. Is that a progression or a digression or is progress mostly a digression? I still throw on the wheel, alter the form gently, often to an oval form, build in the bottom and scrape the walls as thin as I dare. I like the feeling when it is only the surface outside that forms the hollow cavity. Sometimes I throw about five to seven forms and it takes me a week, maybe two, to finish them. The lid, the rim, the gallery, the lid fitting, all are modelled. For larger things, I throw the base, alter it and coil the rest. As my kiln is only 67 cm high, I reach the limit rather quickly. I dream of making pots two metres high.

'I bisque fire now. As my pots become thinner and thinner, they cannot take the wet slip without warping too much. The bisque ware allows me to work out much more complicated decoration. In fact I take longer, or at least as long a time on the decoration, as I take for the forming. Perhaps the painter or the sculptor would be offended by this. A pot, with its inside and outside, with its touching appeal, with its

(Above) Bente Hansen, double container, slip decorated, salt-glazed stoneware, 60 cm/d
Photograph: Finn Rosted

changing surfaces, is a piece of art and wonder unattainable through any other process. After a month or so, I should, if I haven't come across too many other disturbances, be able to fire a salt kiln. Four to six medium pieces, about 25–30 cm high and 12–16 cm broad, and some smaller ones in between. There is always at least one test of a new slip and if no fine pots come from a firing then perhaps there will be a new colour. I prefer to fire one large piece at a time, but this is a risk, although more thrilling. I love to work on a big piece. It fills your eye and your arm — that is a real passion to me.

'The firing is taken slowly. I use two commercial gas burners with the salting ports above them. I make salt boxes in front of each burner to catch the salt. Salting takes two hours approximately. More and more I am seeking the real salt effect on my pots, and I have gradually used more and more salt. Often I refire a piece if I think it doesn't have enough salt on one side. I turn it around in the kiln and often the metamorphosis is astonishing. Of course I'm pleased when the piece succeeds in the first firing, and that does happen.'

Bente Hansen has developed a wide spectrum of colour from white, pale to darker yellow, orange to dark orange, red to red-brown in many shades and characters, to green and black. 'My green is bright and shiny when my luck is out. I make green by covering the black slip with the yellow titanium or cobalt. I also have a slip containing both colouring elements and, depending on the clay base, the green colour has a wide range of shades from cool bluish-green to dull brown, which can be useful in combinations. I'm working with several blacks too, one is matt, another more glossy, without orange-peel effect, just shiny. Together they give an appealing effect which reflects the light.'

(*Above*) Jeff Oestreich, bowl, white slip,
paddled decoration, wood-fired salt-glazed,
10 cm/h, 1989
Photograph: Peter Lee

The pattern or composition which she applies to the bisqued piece is made by spraying on slips, at this stage all delicate grey colours, unlike the fired results. 'It means that I must imagine how the colours are going to work together. I make many small colour sketches before I decide on the decoration for the specific pot. I start with the smallest area, spray that, then cover it with latex and so on for several layers; every new colour has to be covered with latex, and every time the edges have to be refined. When I have finished spraying I remove all the latex and I finally see the whole composition, but still unfired. Often I scratch in thin lines and fill them with another colour. I want the result to be unified, I want the colour combinations to work with the form and the texture in one strong whole expression. I feel that clay is a powerful material and that one must love it before it will reveal itself.'

The balance of important aspects in the art of ceramics — line, weight and proportion plus the unity of form and surface, is apparent in Bente Hansen's work. Her description of techniques and ideas shows that full attention is given at every stage of working, beginning with her love of the materials, using her skills and realising the image she has for her work. Her individual pieces are sold through exhibitions in many countries and she has won awards for her work.

For Jeff Oestreich, a potter from Taylors Falls in Minnesota, USA, salt-glazing offers a way to enhance the forms he is making with a surface that provides interest, and the active possibility of experiment and involvement in processes. First introduced to ceramics at Bermidji State University in northern Minnesota, later he

became an apprentice to Bernard Leach at the St Ives Pottery, UK, for three years from 1969 to 1971. His training has given him an understanding of the functional aspects of pottery, and he has a contemporary attitude to both the visual and utilitarian aspects of the art.

(*Above left*) Jeff Oestreich, beaked pitcher, wood-fired salt-glaze, 22 cm/h
Photograph: Peter Lee

(*Above right*) Jeff Oestreich, teapot, white slip, wood-fired salt-glazed, 1989
Photograph: Peter Lee

Salt-glaze has always been part of Jeff Oestreich's repertoire along with wood firing and glazed wares, although the emphasis has changed from time to time. He writes about the appeal that salt-glaze has for him and why his work is sympathetic to it: 'My attraction to salt-glazed ware began in the late 1960s. Initially it was more the process of firing and salting the kiln that seduced me; and of being totally involved with the kiln, nurturing it along for hours, to be finished with the excitement of throwing in salt, the clouds of smoke and of pulling out hot salt rings. The romance of the process wore off and was followed by a deeper interest in the surfaces possible. By the use of thin slips, more variations in colour are achieved. Blushes and flashing marks provide enough interest and yet do not detract from the form and structure of the pots. Seeing that my consuming interest is in form, salt-glazing does not interfere with this element'.

In an article written by Jim Harris for *American Ceramics 7/3*, Harris talks about the pieces that Jeff Oestreich likes to show in galleries: 'He has selected to refine a limited number of forms that serve as supports for the effects of various clays, glazes, and firing methods. His most recent work is confined to three basic forms: a plate, a vertical vase form, and a teapot. The idea of such concentration returns to an oriental appreciation of the learning value in repetition. In making a hundred similar pots, one understands the differences in individual variations of an established artist'. 'When I work on the same form over time,' says Oestreich, 'it forces me to set a standard which I have to maintain. It also lets me understand the character of the individual piece when I get to see it in comparison with others of the same type; I can work on that character and let it evolve'.

Such concentration on form and its evolution is particularly relevant to the glazing effects of salt or wood firing where flashings of ash or salt vapours do not mask the

(*Above*) Sarah Walton, birdbath, 1989, salt-glazed stoneware

form but add to the unique character of each piece. This single-mindedness on form and surface texture is often combined with self-discipline and dedication on the part of the potter.

English potter Sarah Walton is well known for her strong and direct approach and her uncompromising attitude to all aspects of quality and to the meaning of her work. She grew up in London and studied painting at the Chelsea School of Art, 1960–64, and studio pottery at Harrow School of Art from 1971–73. She worked briefly for David Leach and then Zelda Mowat before setting up her own work-shop at Selmeston, near Lewes in Sussex, in 1975. Her work is salt-glazed at high temperatures.

She writes: 'Chance plays a considerable part in determining the final glaze. The role of the potter is to exploit these occurrences rather than to dictate preconceived effects. Working with random effects is the nature of this exercise. Recent years have seen a reduction in my output, but a greater variety of forms and a wider use of scale. I now work more with handbuilt forms. My techniques have changed with the years; I now try to keep them as simple and as flexible as I can. I started pottery as a thrower and in the last five years have become a handbuilder. What I now throw is subse-quently worked on with handbuilding. Tactile qualities have become increasingly important; so much so that I have wet-ground pieces after firing to simulate the feel

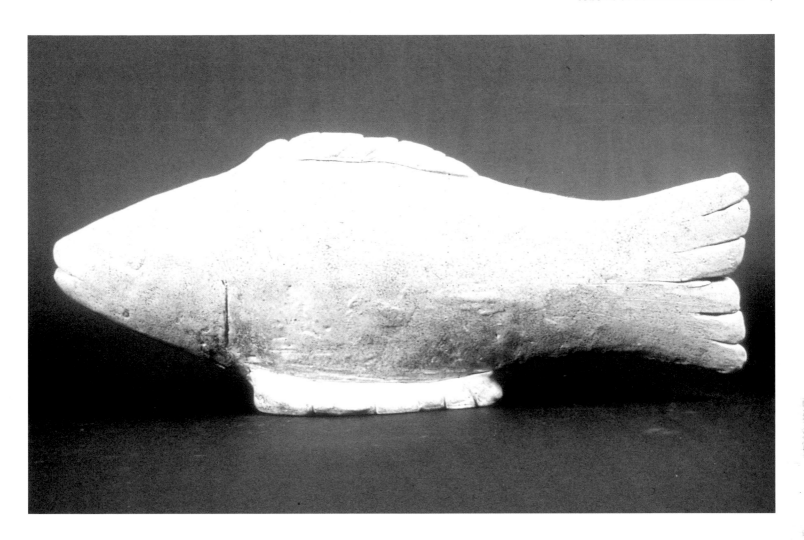

(*Above*) Sarah Walton, *Fish*, stoneware, 57 cm, 1986

of objects that have been eroded by water, weathering or use.

'Throughout the last fifteen years my experimentation has been with form rather than the technique of salt-glazing. Light, which delineates form, is of particular interest to me. The recent forms I have made for outdoors, such as birdbaths, are handbuilt, hollow and with internal supporting walls. To make anything for outdoors is to be working with the play of natural light and a vast surrounding space. That they contain water means that they also reflect light at unexpected moments such as at dusk and during a fall of rain. That I start making them from their centre is apt; I feel my way slowly while developing them, trying to listen more to my feelings than my intellect.

'A blind person must slowly explore the visible world through their sense of touch,' says Sarah Walton. 'In some ways I have worked blind; only after evolving these forms do I recognise what prompted me to make them.'

Working in Perth, Western Australia, Greg Crowe is both a wood firer and salt-glazer. He writes: 'My major areas of interest in pottery are, first, the creation of the form, predominantly on the wheel, and the firing process. I currently use two kilns, both based on the Olsen fast-fire design. One is lined with insulating bricks and is used with small amounts of salt for flashing effects, the other is hard brick lined, has three fireboxes and is used for high-temperature salt-glaze'.

A potter since 1974, alternating between making and teaching, Greg Crowe built his first wood kiln in 1984. He has used wood firing with salt-glazing techniques ever since, always striving to keep the technology of pottery simple. Gradually his pots have moved away from the symmetrical. Now he says, 'I want them to express the

(*Above*) Sarah Walton

(*Right*) Sarah Walton, lidded jars, 13 and
18 cm/h, salt-glazed stoneware, 1987

energy and vitality of the movement of the wheel at the time of making, when the
clay is fresh and crisp. Superficial decoration is becoming less important to me,
rather it is the clay, the throwing process and the firing which give me the surface
that I seek. When I do use glaze, it is a *Shino*-style glaze. The variety of effects that
can be achieved with few glazing materials is proving increasingly fascinating. The
possibilities seem to be widening the more I experiment with the fire'.

Greg Crowe approaches functional ware and individual pieces in a similar manner. He makes groups of pots which explore an idea or a mood, and one idea leads
him to the next. He sells through exhibitions or wholesale to craft shops and galleries. Some of his work is sought by interior decorators. Rather than having a show-
room of work, he says, 'I am more interested in having the space to make more work,
although I admit that, looking at work that has been here for a while, I see more with
each handling. This is especially so with wood-fired and salt-glazed pots'.

The challenges to be faced by a potter involved in salt-glazed production are a
topic of interest and much discussed by them. Australian potter Sandra Lockwood of
Balmoral Village in New South Wales, who also uses wood-fired kilns in conjunction with salt-glaze, writes: 'The main challenges for me, apart from technical difficulties, lie in the area of making authentic and satisfying pots by integrating the form
and decoration, and firing for the total intended effect'.

Sandra Lockwood began potting by undertaking part-time evening classes in
1970 and, in 1977, began as a general assistant at Blackfriars Pottery with Derek
Smith. Her first experience with salt-glazing was as a ceramics student in 1979 at
East Sydney Technical College where she was fascinated and drawn to the process.

'After completing college I bought 5 acres of bushland 160 km south-west
of Sydney. My first priority after moving there was to build a workshop and a salt-
glaze kiln. I lived in a small temporary dwelling and supported myself by part-time
teaching.

'This first kiln was replaced in 1985 by a two-chambered wood-fire kiln utilizing
two Bourry boxes. The first and second chambers are both fired to 1300°C. In the

first chamber, only minimal salt is used in conjunction with stoneware glazes. This produces flashing on the stoneware and porcelain bodies. The second chamber is used for salt-glazing taking about 10 kg of salt per firing. I have experimented with firing times and have found that an overall firing time of 30–36 hours was a good compromise between my levels of endurance and the required results.

'In 1988, I built a 0.7 m³ (24 cu.ft) packing space propane gas-fired kiln. This kiln is partly experimental in design. The kiln chamber and arch are built of refractory insulating brick coated with a high alumina airset mortar. The floor and bagwall are built of high alumina dense brick. This kiln runs on four 37 mm burners, jet size 1.25 mm, at a maximum pressure of 100 kPa.'

Sandra Lockwood unites form and surface in her work in a similar fashion to the way she combines techniques with her ideas. 'It is difficult to separate philosophy and methodology. The way I work is based on what I feel and think, and what I feel and think comes in part from reflecting on the way I work. The simplicity and direct-ness of the relationship of myself to the clay, establishing my own rhythm of work,

(Above) Greg Crowe, bowl, wood-fired salt-glaze, 12.5 cm/d

(*Above*) Sandra Lockwood, platter, wood-fired salt-glaze, 1989

and setting my own deadlines is important to me. Having a peaceful environment and working alone enables me to become absorbed in my work. It is a kind of meditation helping me to become completely focused on what I am doing.

'There are continual difficulties to overcome in order to make good pots and develop as a potter,' she explains. 'Initially, for me, they were how to marshal the physical resources such as clay, kiln and workshop; then there was the challenge of bridging the gap between what was learned in college and the real world. As well as this, there is the continuing adaptation of clays, glazes and firing cycles which are very much part of salt-glazing. I see potting as an ongoing process of refinement of ideas and techniques. This process reflects one's personal development, and a time of no apparent progress is in fact preparation for the next step of development.'

Liking to mix her own clay body to achieve the soft texture and relaxed forms she

desires, Lockwood wants the enjoyment of the making to be evident. 'My approach is to combine all the elements of making, decorating and firing into a state of harmony. Sometimes when I am throwing, joining handles, or decorating, I feel calm, trying things out, and the process feels like play. It is a sense of pure enjoyment, even fun.

(Above) Sandra Lockwood, teaset, wood-fired salt-glaze, 1989

'In life, sometimes, I reach a fleeting clarity of perception regarding something significant to me. So it is with my pots. I occasionally look at one of my pots and have a flash of insight which tells me something about a difficulty I am experiencing with my work, or it shows what the next step should be for me. Such things are beyond logic. I wish my work to show a robust character whilst still retaining balance and subtlety, and a sense of enjoyment. However with potting, as with life, I think effort and self-discipline can bring worthwhile results.'

The successful unity of form and surface in the work of the ceramists in this chapter and their perseverance in achieving this harmony is evident. For them it has been an evolving, gradually developing process; experimenting, then following or discarding ideas until both the working methods have been resolved and a personal artistic style has been achieved. The completeness of expression that salt-glaze is able to attain, where every part of the work comes together to form a balanced personal statement, is an ideal sought in all forms of art.

7.

A LIFETIME OF EXPERIMENTS

For some ceramists, the ideal is always elusive. Even more than that, the wish to discover, to try out ideas, to search and research is, in itself, a fulfilling pursuit. In salt-glazed ceramics, a potter can have an endless involvement with ideas and techniques. Each firing can be an exploration of new theories on materials, processes and aesthetic expression.

Claude Albana Presset from Geneva, Switzerland, is a researcher and experimenter into ideas that can be portrayed through clay and salt-glaze. She first discovered the beauty of salt-glazed wood firing at the traditional kilns at St Armand and La Borne in France in 1954. Here was a medium which she felt she could explore, combining all the different values which she believes work together in ceramics. After studying oil painting at the Ecole des Beaux-Arts in Geneva, she trained as a potter, first in Switzerland and later, in 1960, with the Arakawa family in Japan. From her Japanese masters she felt she learned an attitude towards clay and wood firing which allows her to research her work in a way that 'reduces the personal ego and keeps one in the limits of time and space while transporting one's perception to another spiritual dimension'. Continuing since then to work as a teacher and practising potter, she has built wood-fired kilns, concentrating on salt-glaze.

She writes: 'With salt-glaze it is possible for me to achieve a brocade-like effect with slips which, in conjunction with carving, make patterns to catch the light; over the darkness and thickness of different coloured clays, the varied transparency of the surface glaze is capable of absorbing and reflecting life, like a waterfall over rock. Salt-glaze gives both richness and simplicity to the clay surface and I believe there is much yet to be explored using techniques over and under the glaze layer'.

Claude Presset says she is happy in the role of explorer. 'Sometimes it is possible,' she says, 'to see these effects on my pieces, these special qualities, but only in detail, not on the whole shape. The effects seem to come from inside the clay, appearing and disappearing at the same time, and if we are caught by the surface we are drawn into the inside perception of clay. While many ceramic techniques can offer this possibility, only salt-glaze has the perfection of this. Salt-glaze can achieve a realistic, materialist and vivid contrast. Using porcelain, the choice of the material will give whiteness, half the work is done, and a contrast of black ink will give distance and heaviness by opposition. I seek opposition in all things, balancing the light and heavy, light and dark, in colour, in forms, in the clay itself. Black painting on white clay gives weight to the object, holes through dark clay introduce light.

(*Above*) Claude Albana Presset, *Column* (detail), black manganese stoneware, salt-glazed in wood kiln, 1.55 m/h

(*Facing page*) Richard Launder, dish with curved base and holes, residual salt, 1300°C, 29 cm/d, 1987

(Right) Claude-Albana Presset, *Stele*, salt-glazed in wood kiln

'Whether I am making a bowl or a cup, or a column of bricks, the same idea applies. The waterfall expresses the inside of the mountain. People looking at my work can be attracted by one or other of these contrasting aspects, or even respond to them both, never taking the middle line. Salt-glaze is not merely a surface but is part of the clay body itself. It is, for me, not a hard reflecting surface but lighter, with many variations. There is even more possibility to achieve these variations with wood kilns. Beside the physical body there is a spiritual body standing beside us, felt but not seen.'

The work of Richard Launder reflects his enquiries into concepts and techniques. He too is an explorer. He writes: 'Salt-glaze is a relatively unresearched area and,

being inquisitive, I enjoy the intellectual challenge that this gives to the mind, emotions and intuition. Improvisation, living on your wits, learning to trust my intuition; these are important in my life everyday, as a maker and a teacher. I must assess and criticise, decide what to keep, change, find where to begin and go on. Over a period choices are made and a familiarity with a particular palette develops with attitude, material, construction, kiln work, and a response to concept and theme'.

(Above) Claude Albana Presset, *Steles*, stamped and press moulded stoneware, white slip, celadon glaze, salt-glazed in wood kiln

One of the options which Launder has chosen to explore is the use of salt-glaze at earthenware temperatures. In an article on the subject for *Ceramic Review 70* in 1981, he said that he believes there is potential for rich and varied effects in the 1100° to 1120°C range. 'Earthenware clays,' he stated, 'have a high content of iron and calcium which is generally not desirable in salt-glaze, but can be used to advantage'. He has worked with light and dark-coloured clay bodies and found the lighter ones to be more useful, as they can take more heat and salt. Experimenting with different clays and slip colours and glazes, he is able to achieve a strong texture from the salt-glaze. Further experimentation with reduction and oxidisation of the kiln atmosphere can result in specific and unique effects with the advantages of using less fuel, less labour time and the possibility of using cheaper clays.

Launder studied ceramics at the West Surrey College of Art and Design in Farnham, England, from 1972 to 1976, and has worked and exhibited in England, Greece and Norway. He is currently Associate Professor of Ceramics at the National College of Art, Craft and Design in Bergin, Norway, and for half of the year he lives in England, working at Upper Hale, Farnham.

(Above) Richard Launder, teapot with diamond sprigs, porcelain, light salting, 1300°C, 12 cm/h, 1985

Writing on Launder's work in the *Ceramic Series* for the Aberystwth Art Centre (No.1, 1984), Geoffrey Fuller said: 'One must admire the vigour of the throwing, the astonishing range of colour that moves from a deep, masculine blue and brown to gently soft feminine colours that no other salt-glaze potter seems to be exploring in this way, at this time. The pots spring from a knowledge of and care for centuries of pottery tradition. Shape and colour is used in a total clay context'.

Concept and purpose is another area for exploration which engages Launder. In an essay entitled 'The Aesthetic Vessel', he probed some of the problems facing the ceramic artist today: 'Due to industrialisation and the development of mass markets,' he wrote, 'the domestic and applied art scene during the last hundred years has been concerned with purely decorative mantlepiece-dust-collectors. The ritualistic and symbolic aspects have been debased and the objects have lost their meaning. Traditional crafts have a narrow aesthetic language. Contemporary crafts embrace a

much wider range of concepts, expressions, attitudes and philosophies. There is a paradox here. We need to go deeply into the vast art and craft heritage in order to stay alive but, on the other hand, there is a hunger for a wider aesthetic vocabulary. The balance of function versus expression is changing towards abstract and symbolic forms and the familiar components of the piece are changed by the context, releasing latent meanings from familiar functional and historical pot types'.

(*Above*) Richard Launder, tea bowl, granite grog, residual salt, 1300°C, 11 cm/h, 1987

Beauty, utility and sculptural quality have been vital factors in the search for self-expression made by artists throughout the history of ceramics. 'Peter Meanley's pots are both sculptural and functional, they are uniform yet have individuality,' wrote sculptor John Kindness reviewing Meanley's salt-glazed teapots exhibited at the Crafts Council Gallery, Ireland, in 1988, drawing attention to the sculptural essentials in pottery.

Peter Meanley has always worked to a theme. In the early seventies he made flat

(*Above*) Richard Launder

irons, blow torches and rocking forms. Later, it was clocks and by the mid-eighties it was bowls. For him, the idea comes first; he sees and feels something that becomes important to him to translate and make. With the bowls, he developed a glaze and perhaps eighty slips that were reliable and he was able to produce illusionary effects through line, tone and colour. Drawn to make domestic pots, he discovered, or as he says, re-discovered eighteenth-century English teapots.

'I can still recall the feeling, the intense power and presence of these pieces: Whieldon, Astbury, Wedgwood, agate, tortoiseshell and salt. The forms were light but they had a hardness of edge and a spring of power. The relationship between the parts was sure with strong terminating points. The quality of the brush and lyrical use of line and pattern was uninhibited and expressive. Often the same processes were used that we use today. All I wanted to do was make teapots, explore teapots, differ-ent forms, different combinations. All my work before has logically led to this; the teapot is supreme with the small attachments to the main form, the way they relate, the spaces left in between and the vitality of the whole. I felt that salt-glaze, as used by those eighteenth-century potters whose work I so wanted to emulate, would give me total freedom and yet consistent mouth-watering fired quality and enable me to realise the fantastic ideas that I wanted to make.'

Meanley says he is not a ceramic chemist who can logically solve problems of clay and glaze, but he is reasonably practical. Wanting a white, vitreous hard clay body to

use with colour, he tested various clays he had been using for his previous work. He painted the interior of his existing kiln with a salt resistant wash: 5 alumina hydrate, 2.5 molochite 130#, 2.5 china clay, 1 silicon carbide 130#. This resist worked well on the hot face insulating bricks, the shelves, bagwalls, props and inside the chimney base. Eighteen firings later the kiln is still holding well.

'In the first firing I used 7.5 kg of salt but I have brought this down to about half this amount in recent firings. I always log the kiln so that I know the pack, the wind speed, the amount of propane in the tank, the level of the bottom shelf in relation to the bagwall; the damper settings can be critical as well as the rate of climb. I use wood from railway sleepers to augment the reduction and I keep a record of how much I use each time. I put in two sets of cones (Orton), top and bottom, 06 to start reduction, 3 to commence salting, 5, 8, 9 to complete salting and commence oxidising, and 10. I always put in four or five test rings so that I know what is happening throughout the salting. This is important for the salting but even more so for the reduction.

'My kiln is small, 0.1 m³ (4 cu.ft), but perfect for up to twelve teapots, and consistently even in terms of temperature and salt. There are two burners with long flame

(*Above*) Peter Meanley, teapot, slip decorated, salt-glazed

(Above) Peter Meanley

length. The flame is deflected by a small brick in both fireboxes in different positions to throw up the heat; this ensures equal temperature and equal salting from front to back. I drop the salt (slightly damp so that it stays firm) on to the deflecting brick, in small amounts continuously from cone 3 when the still non-vitreous body allows the silica in the clay to combine with the sodium in the salt more easily. The firing usually lasts eleven hours. When cone 10 is flat and I have soaked the kiln in oxidising conditions for one hour or less, I rapidly cool the kiln for a few minutes with everything open and the burners turned off to get rid of the residual salt. At about 1100°C I close the secondary air ports but leave the damper slightly open as salt vapour is still leaving the chimney. I believe this stops scumming or dulling of the salt-glaze surface.

'Three years on with teapots,' writes Peter Meanley, 'I am still excited by them. I made an early decision not to slip-cast but I am prepared to pressmould, extrude, pull, throw and turn, cut and reassemble. I make enough "bits" for about four teapots: feet, belly, shoulder, spouts, handles, lids, knobs. I push clay out whilst throwing through jigs and dies to achieve a vitality which is non-mechanical. I use roulettes on handles after pulling them, then cut, reassemble and add extra bits. After bisque firing to about 1020°C in an electric kiln I play with slips (50 ball clay, 50 china clay, 10 oxide or stain) and am using hot wax in conjunction with the slips to build up layers, often with stamped slip patterns as well. I am prepared to re-fire as many times as need be, and am considering the use of enamel over lightly salted oxidised surfaces. With my commitment to the University of Ulster, Belfast [Meanley is currently senior lecturer in ceramics in the Fine Craft Design Bachelor of Arts degree course], I have the opportunity to seek quality without worrying unduly about quantity or accounts, as I realise that sometimes the business of surviving blunts the ability to experiment and to reflect'.

(*Above*) Peter Meanley, teapot, slip decorated, salt-glazed

Meanley is a true experimenter, and generous with his information. His ideas chase one another in rapid succession, and he follows them in a practical and direct way. The small kiln allows a constant turnover of experiments involving form and surface quality, all pursued in a state of energetic and spirited achievement.

For John Chalke of Calgary in Canada, experimentation has been a hallmark of his career as a ceramic artist. As he wrote in an article for *Ceramic Review 116* in 1989, his curiosity has taken him through all the firing techniques, clay research and glaze studies, sometimes going back to revamp and reshape earlier tests. One of his experiments was with soda-vapour, using sodium carbonate as a salt substitute.

'This process involves spraying caustic, but environmentally agreeable sodium carbonate in an aqueous solution through two spraying ports directly above the burners on either side of the kiln door. For my 0.85 m³ (30 cu.ft) downdraught natural gas kiln I find the mix of 1 kg (2 lb) of sodium carbonate to 4 L (1 gal) of hot water suitable. I let it dissolve for 15 minutes and then sieve the solution through a 100-mesh screen into a metal weed sprayer which is then pumped to pressure by hand. My top temperature is cone 6 so I begin to introduce vapour when cone 5 is starting to soften. After damper adjustment I inject 20–30 short, two-second bursts in through one port and then move over to the other side of the door. This pausing style

(Right) John Chalke, plate, coarse felspar
additions, slip trailing, 45 cm/d

of spraying was to allow steam to disperse and also to add a ritualistic timing. The
spraying, once begun, continues from one port to another for approximately 30 minutes. I let cone 6 go down in oxidation (15–20 minutes) and then shut off. The pro-
cedure is the same for earthenware firings at cone 05 or so, except that I use about
30% less sodium carbonate. Red earthenware clay shows the effect of damper reduction graphically and the ware varies from bright orange in some parts to a darker
chocolate brown.

'The clear advantages of sodium carbonate are less wear and tear on the kiln and
kiln furniture, and one's urban environment. Costs, in spite of the spraying equipment, are cheaper and kiln temperature rise is hardly affected. I believe the glaze surface itself is actually preferable to that of sodium chloride on my clay body which
contains some nepheline syenite, being softer underneath overhanging surfaces and
generally giving a gentler range of colour from blue greys to orange; there is a
reminder of the vapour's soda origins.'

Patricia Ainslie, Curator of Art, Glenbow Museum, Canada, reviewing the 1989
show 'Vessels Fe' at Glenbow Museum in *Contact* magazine (No.76), said of Chalke's
work: 'John Chalke is a collector of experiences and his work carries residual and
overlaid memories of places and events: life-layers of personal history. He makes
objects based on functional forms but their use is ambiguous or denied: plates and
bowls are hung on the wall rather than laid on the table, a skillet has only vestiges of
a handle, and surfaces are highly worked and textured. The glazing of these objects is
multi-layered, added to repeatedly, often over the span of several years'.

Again in *Contact* (No.77), Ron Moppet and Mary-Beth Lavoilette, reviewing the
'From Clay Exhibition', reinforced the unusual and exploratory approach of this
artist: 'The experimental, unorthodox surface is a Chalke trademark and in a series

of recent wall plates, each with a crude handle attached to the edge, it reaches a kind of mad, molten crescendo. Cracked, blistered and full of bumps, these plates and their wonderful surfaces take on an extra-terrestrial presence that is both visually seductive and suggestive'.

(Above) John Chalke, *Cornering*, soda vapour glazed, 28 cm/h

For the salt-glaze ceramist every firing has the potential for experiment. This has to do with the number of variations possible. Even though the clay body, the slips and glazes are well tested and the kiln is fired following the log of previous successful firings, differing amounts of salt volatilising at slightly different temperatures can give unexpected results. If the stacking arrangement of the pieces in the kiln is altered in a minor way or if the changes in the atmosphere in the kiln follow a slightly different pattern, there can be surprises on opening the kiln. How the kiln is cooled, quickly or slowly, and whether cooled in reduction or oxidisation can also be factors that give the ceramist trouble or delight. After examining the results of each kiln load the salt-glaze ceramist has a host of new theories. Many of the ceramists who contributed to this book said they feel there is still so much to discover that, even after many years, they are just beginning to understand and use these variables to advantage. It is the possibilities, the combinations and the myriad of new options imagined when looking closely at salt-glazed pieces, that beckon a ceramist into a lifetime of experiment.

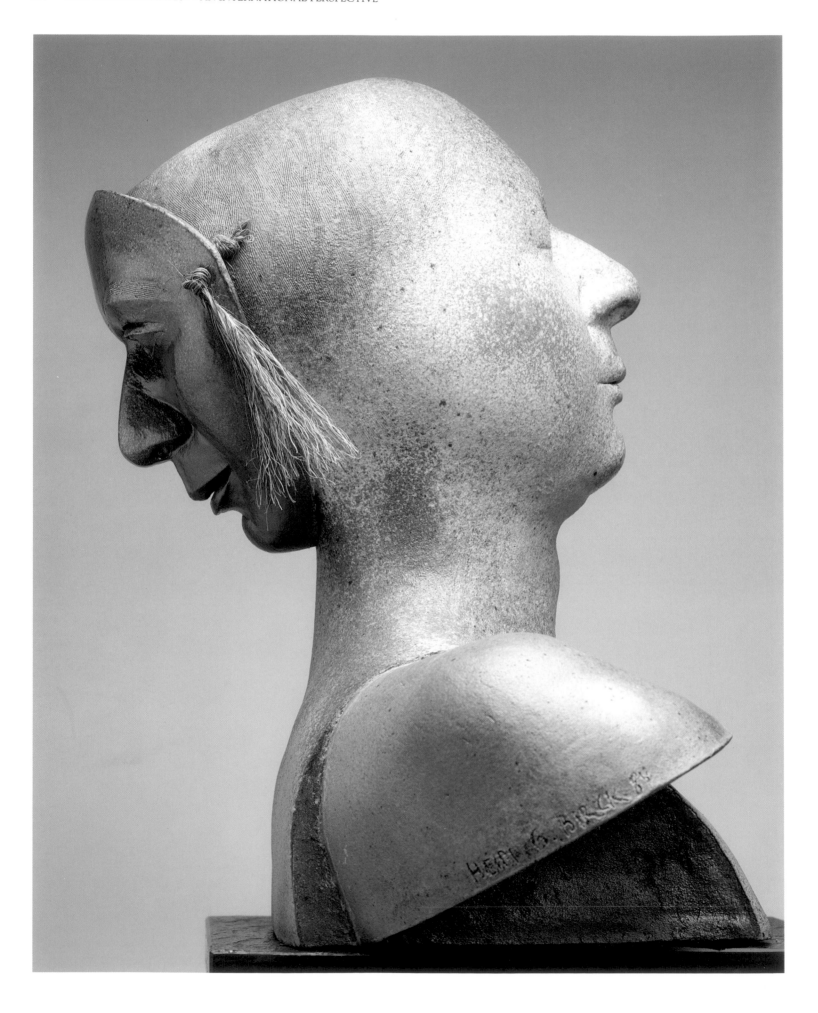

8.

THE SALT FACTOR IN CLAY SCULPTURE

T he nature of clay invites manipulation. Some of the earliest artefacts ever made, small figures of humans and animals, were modelled from clay as long ago as the Ice Age. The ability of clay to take on any form in response to the artist's intention, and then be rendered permanent through the firing, has given us a knowledge of civilisations since that time. We are able to see a sequential evolution of ideas. Today, the field of ceramic sculpture is one of the most progressive areas of applied art. Artists are using the potential of the material and processes to express perceptions of mass and scale, human frailty and hope, as well commenting on our social circumstances and environment.

One of the vital considerations for a ceramic sculptor is to use an appropriate surface for his or her work, especially when the form of the work is of paramount importance. Salt-glaze, with its overall texture, has an ability to take a sheen of controllable brightness, to reveal surface marking and to exaggerate texture by highlighting edges or softening planes. For Maria Geszler, a sculptor from Szombathely, Hungary, who uses a process of silk screening photographs on to porcelain slabs before shaping them to her requirements, salt-glaze is the perfect resolution to complete her work. 'For me, salt-glaze is not technology, salt-glaze is content,' she explains. 'I feel I have at last found the technique of softening the hard lines of the screened photograph by this method. Salt-glaze envelopes the porcelain mass, changing the colour of the cobalt oxide to pure silver. I like the white grey clay with cobalt painting, and after firing the silvery salting of the cobalt gives my porcelain windows with their silk-screen prints a surrealist content.

'For years I have endeavoured to bend, fold and shred porcelain. This material, pure and white, sensitive yet malleable, lends itself to these expressive effects. My initial experience began with folding porcelain clay to resemble windows and curtains. I realised that an image screened and transferred to the wet surface of porcelain highlighted attitudes and profundities. If the vestment of the built-up figure is a screened image, the folds follow the movement of the figure. This offered me new possibilities for dramatic expression.'

Without a salt-glaze tradition in Hungary, Maria Geszler turned to German pots of the Rhineland for her model. 'My small native country is in the middle of the east European basin, surrounded by mountains and inhabited by Hungarians. These are only modest hills, mild slopes; Hungary is more renowned for its sunny, sandy and extensive plains. We have a wide variety of coloured clays: grey, ochre, yellow, red;

(Above) Maria Geszler, *Grass figure,* porcelain with paper print, salt-glazed in wood kiln, 21 cm/h, 1300°C, 1984

(Facing page) Heidi Guthmann Birck, *Mask and dream,* salt-glazed stoneware, life-size

(*Above*) Maria Geszler

(*Right*) Maria Geszler, *Solitary bench*, porcelain and grog, silkscreened salt-glazed, 1310°C, 27 x 25 x 1.5 cm, 1986

but they are for low-firing temperatures, 960°C. During our long history of rich folk-lore, including pottery, we have developed dishes for everyday use and objects for decoration thrown by hand on a wheel, with an exuberant decoration painted with coloured clay or covered by lead glaze and fired in a wood-firing kiln.'

Maria Geszler studied the history of salt-glazed objects at the Academy of Applied Arts in Budapest and had the opportunity to see salt-glazed pieces in museums abroad and in her own country. She was deeply impressed by medieval German jugs and cups 'made of a white-grey basic mass decorated with cobalt spread by a salt atmosphere'.

When, in 1978, Janos Probstner founded the International Experimental Studio Ceramics in the Hungarian town of Kecskemét, ceramic artists were able to study and experiment with materials and technologies inaccessible earlier. It was here that Maria Geszler had the opportunity to work on porcelain, to work a silk screen and to fire a salt kiln. Since her school-girl days she had passionately photographed land-scapes, seeking hidden messages. She visited these places in various seasons and moods to catch the appropriate moment, in some cases over 10 years, for the desired light, or exaggerated richness of nature. Film, photography and computer graphics revolutionised applied arts for her, and Andy Warhol's photo-screen graphics and surrealism influenced her own 'internal inclinations, emotions and dreams. I am attracted by mystery, by the untouchable, by symbols and by enigma,' she says.

Geszler describes her techniques: 'I use a 1360°C porcelain body or grog (chamotte)

(*Above*) Maria Geszler, *Autumn scenery*,
porcelain with silk-screen print,
41 x 43 x 19 cm, 1986

(Above) Emidio Galassi, *Untitled*, refractory clay, wood, 350 x 110 x 30 cm, 1989

and porcelain mixture. My screen pastes are made of pure metal oxides and poly-ethylene-glycol. The objects are rarely glazed or, if so, I use a salt-reactive glaze, sensitive to wood ash. A glaze which is matt in texture from 1250° to 1300°C is composed of 24 dolomite, 4 whiting, 48 felspar, 24 kaolin and 5–8% cobalt carbonate. An orange-yellow matt glaze for the same temperature range is 24 whiting, 61 felspar, 15 kaolin, with 5–10% rutile.

'If possible I avoid bisque firing. I am convinced that with single firing the salt is taken into the material more deeply. I begin the firing slowly, instinctively adapting myself to the weather, the kiln, the quality of the wood etc. After having attained 800°C the pace of the firing increases until 1300°C is reached generally throughout the kiln. I salt three times with sodium chloride, at 1250°, 1280° and 1300°C respectively and calculate for the ash. I reduce the kiln strongly. I fire over 10 to 12 hours and prefer a small kiln that I can manage by myself or with little help. I always fire my works myself. My objects are my life.'

To describe the works of Maria Geszler, their sincerity, their purposefulness, the whimsical yet determined stance of the figurative pieces, is to describe the artist herself. The work is a reflection of her spiritual dimensions and her commitment to her art. There is a depth to her work and behind the imagery one can sense her dedication to express values that are important to her, her country and to all people.

Emidio Galassi, from Faenza, Italy, is a sculptor working with refractory clay forms to produce architecturally influenced free-standing forms. He has won a number of prestigious awards for his work which has been exhibited in many countries. He writes on his experience of salt-glazed stoneware: 'I like to use both white and coloured refractory clays, taking it to 1200°C. My downdraught kiln is fired with gas. I introduce the salt, in proportion to the dimension of the kiln, at 1100°C and then

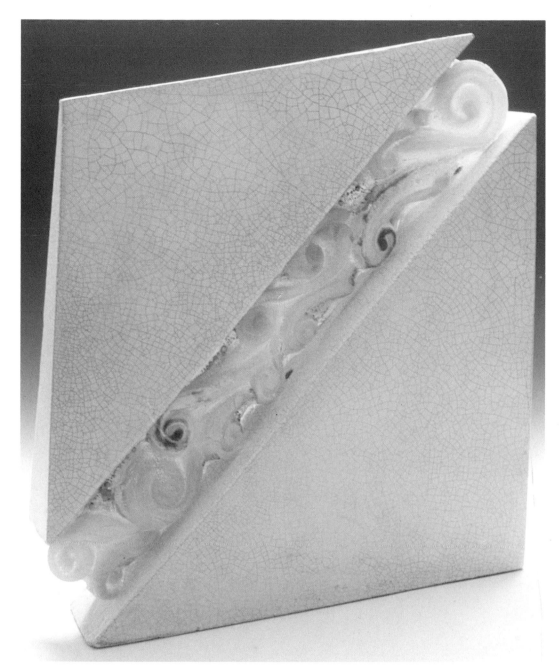

(Left) Silvia Ferrer, *Sculpture*, glass and ceramic

again at 1180°C. Sculptural pieces are slipped and glazed making the surfaces more rounded. The experience of salt-glaze is a positive one, giving great possibilities for sculptural application'.

In a catalogue introduction by Diva Ponti and Josune Ruiz Infante for an exhibition of Galassi's work held in Spain in 1988, reference was made to the artist's role as observer and experimenter: 'Emidio Galassi develops to an extraordinary degree the capacity, not only to communicate knowledge, but to transmit a renewed force of enthusiasm and a desire for contrast and investigation'. Galassi is described as 'fleeing from that inner contentment that blocks the flow of creativity,' rather to seek that 'moment of peak creative energy in which the unexpected bursts forth in the search for new paths'.

Cecilia Nalbandian and Silvia Ferrer are two South American ceramists, currently working in Italy, who have chosen salt-glaze to achieve the 'natural effects' possible on their sculptural work. Silvia Ferrer was born in Mendora, Argentina, and obtained her diploma of art from the National University of Art of Cuyo in

(Above) Cecilia Nalbandian, *Project for a monument*, salt-glazed stoneware

Argentina. She became a teacher, took part in group and individual exhibitions and was awarded prizes for her work. She then won scholarships to study in Italy. Incorporating glass as well as clay in her sculptural work she says she is able 'to achieve the contrast between the warm reflection of light and coolness of geometry, and so express the presence of humanity'. Cecilia Nalbandian was born in Montevideo, Uraguay. She has diplomas of design and ceramics as well as interior decoration. She also has won awards to study further in Italy. In her work she is interested in the play of light and shadow in organic carved forms. Together, these ceramists are exploring the techniques of salt-glaze to complete their sculptural work.

With a preference for the results they obtain with high-temperature salt-glaze, they have experimented widely with clays and engobes. Only a thin surface of salt is desired but they accept that the technique is not altogether controllable. One particular practice they have found to be useful is to place the sculpture inside a closed box, or saggar, inside the kiln itself. A mixture of 67 marine salt, 28 calcium carbonate and 5 sand is painted on the inside of the closed box. They report: 'At 1200°C the salt will volatilise, and be attracted to the silica in the clay body of the sculpture. When putting the sculptures inside the box, we need to see that the salt can easily

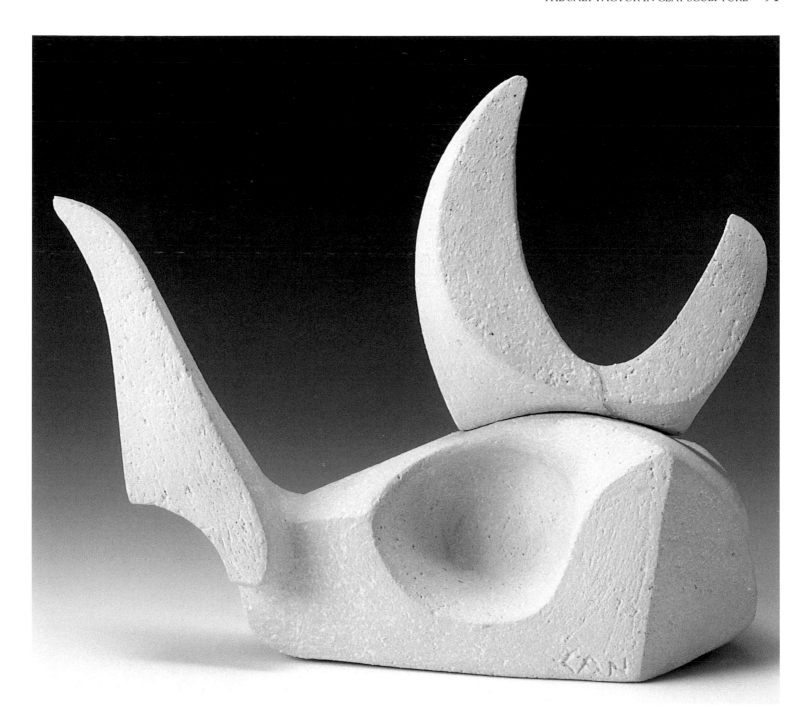

(*Above*) Cecilia Nalbandian, *Curved form*, salt-glazed

reach all surfaces and there must be sufficient space between each piece to achieve an even coating. It is important to control the cooling of the kiln after it has reached the maximum temperature. If the temperature falls slowly, the resultant colour is a typical brown, and if it falls quickly then the colour will be grey-green or almost black'.

Peteris Martinsons of Riga, Latvia, takes the opportunity to work with salt-glaze whenever he can as he feels that it is a totally suitable surface for his sculptural pieces. A recent series of works which he calls *Boxes* relates to what he perceives to be the sad condition of people in today's society. 'They are all hemmed into cages,' he says, 'cages like boxes that they have made for themselves and into which they are now trapped'. A ceramist and teacher, Martinsons has made many large architectural pieces, in particular, free-standing works for placing in courtyards. However, as he declares: 'My work is intended to be a statement of art, not a political statement'.

(Right) Peteris Martinsons

Aruni Constantanidi is a ceramic artist and also a teacher. She was born in Greece and trained at Isleworth Polytechnic and Hammersmith College of Art and Design in London and then at the West Surrey College of Art and Design, Farnham, England. She has worked in England, Greece and Scandinavia, exhibiting her work regularly in these countries.

In an introduction to Constantanidi's work at the Gallery Bossky in Copenhagen, Denmark, Richard Launder drew attention to her questioning character and her search for personal imagery. He wrote: 'She has stayed clear of movements and styles, choosing a hard, slow path of independence, keeping her artistic integrity and direction intact against current tides or trends. The images of her forms state her personal search, at times with passion, at times languid or calm. They draw one in. The concept is the source of her work, the manner of execution and the materials she uses follow'.

Constantanidi uses clay in an attempt to convey her emotions, fears and hopes, her knowledge of despair and her search for happiness. Her research into the under-

(Left) Aruni Constandinidi, *Tama to Aphrodite*, slab-built, low salt firing

standing of beauty and into the softness and vulnerability of clay to express her feelings in a tangible form, has occupied her creative life. Her concerns include the attitude and responsibility of the artist to communicate and keep an open investigating mind. Working in low-salt, post-smoked techniques on her sculptural ceramics, in the temperature range of 1000° to 1100°C, she experiments with porcelain clays, slips, underglaze colours and fuming chlorides and uses a blow torch on the already fired surface to stabilise enamel and gold lustres and obtain post-firing oxidation.

Born in Germany, Heidi Guthmann Birck established her first workshop in 1965 in Copenhagen after working in French and German studios since 1959. She began

(Above) Heidi Guthmann Birck

exhibiting in 1966, and has continued a programme of exhibitions in many countries, mostly in collaboration with her husband, Aage Birck. Working with stoneware clays, Guthmann Birck's sculptured heads portray her dreams, visions and fears through the attitudinal expressions of the head and face. 'My intention is to show the reality behind the visible,' she wrote for an exhibition at the Museum for Moderne Keramik in Deidersheim, West Germany in 1882. 'I use symbols and realistic elements to present the flow of emotions and state of mind in the language of the face.'

Heidi Guthmann Birck believes that salt-glaze is the preferred finish for her ceramic sculptures. She writes about her experience with salt-glaze: 'The first salt firing in a test kiln created various interesting colours on the surface of my small wall sculptures. At that time, in the mid-seventies, these wall sculptures represented human faces and I used texture and different slips combined with salt-glaze to underline structural effects. Later on, with more sculptural work, I found that to use only one glaze and one slip combined in different layers, or separately, made it possible for me, with the influence of salt-glaze, to get a colour scale from dark grey over light grey-pink to pink-orange. One particular grey matt glaze that is useful in combination with slips is based on 25 whiting, 40 felspar, 35 kaolin with 5% of rutile added. The salt gives the surface a touch of *wetness* like a wet stone. This character of wet stone I try to achieve every time I work with salt, to give the colours greater depth.

'Aage Birck and I share our firings in the salt kiln but I prefer to fire the kiln in a residual salt atmosphere. This means firing the kiln after a firing that has been heavily salted. The materials of the kiln walls have absorbed enough salt for one more firing, without adding more salt. This residual salt atmosphere will release enough vapour to give my work the *wet* touch. In a normal salt firing the result can be sometimes too shiny, and over salted areas on the nose and forehead of my figures can be detrimental to the final artistic expression.'

Wishing to examine the tensions evident in the human face, between the outward facial expression and the inward character, Guthmann Birck's figurative work often deals with opposites. Slabs of clay are cut away to reveal a hidden countenance or a head has two faces with contrasting expressions. Faces are masked, in part or total; some are cut linearly to reveal a second person somewhat out of synchronisation with the first.

Writing on her techniques for *Ceramics Monthly* in April 1986, she explained how she makes her ceramic heads: 'Usually the work is started with a clay prototype of a female or male head. A two-part plaster mould (front and back) is cast from this head. With the two moulds positioned horizontally, 6 mm thick slabs are firmly pressed into each. After the clay has stiffened somewhat, the edges are trimmed and the moulds inverted on a flat board. Released from the mould, the back of the head section is covered with plastic, then returned to the horizontal mould for support. Next, the clay edges are rubbed with slip and the front section joined to the back. After removal from the mould, specific modelling of the head begins'.

The slight sheen to the work from residual salt, as in the case of Heidi Guthmann Birck's work, or the heavier texture of a typical salted surface, are two of the possibilities using salt to provide a surface for ceramic sculpture. Another is using the reactions of the salt in a closed saggar, as described by Silvia Ferrer and Cecilia Nalbandian, to provide a localised salted surface. In all cases the salt-glaze layer is capable of giving a permanent, water-proof and aesthetically satisfying finish to the work. Before opening the kiln after the firing, the ceramist will have an idea in mind as to the expected result. The actual result could be quite different.

In what could be called post-firing assessment, the artist, confronting the piece as

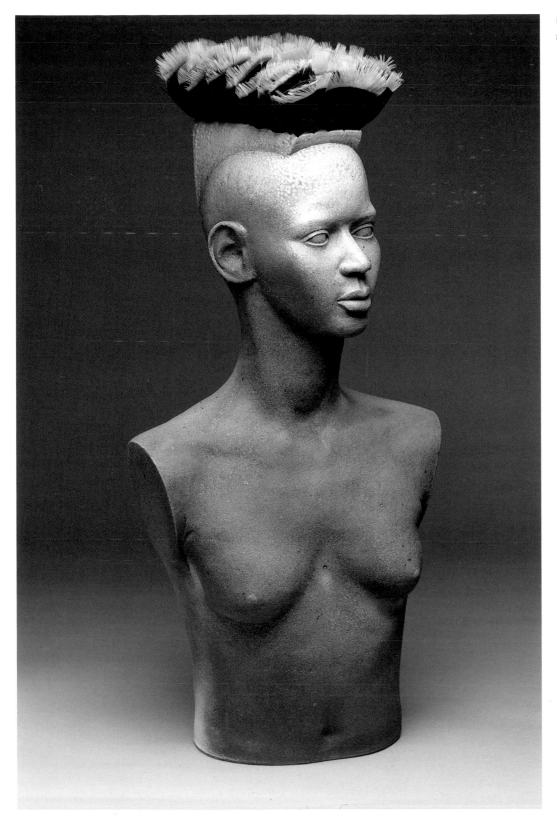

(*Left*) Heidi Guthmann Birck, *Nykredit*, salt-glazed stoneware, 60 cm/h

it comes from the kiln, must make an evaluation. The final responsibility of deciding whether the work is good rests with the maker. In domestic ware, the decision, in terms of colour and depth of glaze, relates to its intended use. In sculptural work, with its quite different objectives, contrasts of rough and smooth texture, of colour variations or movement such as warping, will all need to be individually and critically assessed.

9.

THE FULL FLAVOUR OF SALT

Anyone familiar with the appearance of the inside of a well-used salt-glazing kiln will recognise that tremendous forces have here been at work. The salt has melted the refractory brick lining, causing the hot face to flow down the sides of the walls; glassy stalactites hang from the ceiling waiting for the next firing so that they can bless or ruin, with large glossy green tears, those pots that lie underneath; the fireboxes and bagwalls resemble volcanic rock formations and the walls of the kiln have taken on a convexity as if blown out by violent internal pressure. This description fits an old kiln fired for many years by the author, Janet Mansfield of Sydney, Australia. Like the axe which served its owner well over 30 years with five new handles and three new heads, this kiln has been revived with two new arches, one new side wall and several firebox rebuilds over its 15 years of use.

Janet Mansfield has been involved in the salt-glaze process for many years. In her work she wants to see the evidence of her ideas and the full revelation of the processes of salt-glaze. The pots should depict all the turbulent forces of the firing and look as though 'they have survived the heat and salt vapours and been enriched by them'. She enjoys all the activities involved in making pots from the physical effort of preparing the materials, making and firing the work and the research into how the materials will react together, to the important consideration of aesthetic ideas and their development. 'It is from the activity that challenges come,' she says, 'that new ideas are explored. This personal, conscious input should be visible. So much that happens in the salt-glaze process, although not under strict control, can be anticipated and encouraged by the potter. The materials, the whole style of working, throwing, turning, the forms and their attachments, the decoration and the kiln firing methods should all make a coherent statement of intent'.

Since 1975 most of her research has been devoted to salt-glaze practice. She uses the clays and minerals from the local area, blending them slightly differently for darker or lighter results, using more fluxing materials or, conversely, a higher percentage of gritty unwashed clays, depending on the intended firing and style of work. The clay body, she says, is the major consideration in determining colour, texture and throwing quality. She makes a range of vessels; in particular, she likes to make large jars in three or more pieces and assembled while still soft. All the forms are made with the aesthetic potential of salt-glaze in mind. When a piece is leather hard she often uses a shaped tool to scratch a pattern or a carved wooden paddle for texture.

(*Above*) Janet Mansfield, jar, salt-glazed, wood-fired, 1300°C, 48 cm/h

(*Facing page*) Christopher Staley, *Auger*, porcelain, salt-glaze over copper and rutile glazes, cone 8, 65 cm/h, 1988

(*Above*) Janet Mansfield, covered jar, salt-glazed, wood-fired, 1300°C

The use of handles is a characteristic of Mansfield's work. She says: 'Both the shapes and the patterning of the attachments catch the salt vapours and are accentuated by the salt-glaze. I use slips, again a mixture of local materials to give more contrast to the surface. Every firing is an experiment with so many variables possible. The firing takes one's complete involvement, especially using wood-firing kilns. I fire to over 1300°C to vitrify the clay body and this ensures that the salt sits richly on the surface. I usually commence salting when cone 9 is half down, taking two hours to throw in enough salt to coat the ware, about 15 kg per m³ (1 lb per cu.ft) of the kiln capacity. While some potters wrap dampened salt into parcels for throwing into the fireboxes, others blow it in, or fill small long-handled shovels or pieces of angle iron which are then pushed into the kiln and turned over to drop the salt, I have always found that the simplest and most effective way is to throw the dry salt, from a cup, directly into the firebox. The important thing is to throw the salt where the kiln is the hottest. Then the vaporisation will be rapid and forceful, especially if there is a small pool of molten salt already there from previous charges of salt. I use draw trials or test rings to determine the amount of salt build up. When I am satisfied that

(*Above*) Janet Mansfield, *Flower Vase*,
fired in anagama kiln for 3 days, 37 cm/h

(*Left*) Janet Mansfield, *Jar*, stoneware,
wood fired salt-glaze, incised pattern
through slip, 45 cm/h

enough salt has been used, I clam the kiln up tightly.

'The brown, red and yellow colours come, I believe, from the slow cooling of the iron crystals melted in the glaze layer. My kilns are all made from a local white building brick. These are heavy, dense bricks and while they take more fuel to heat, they retain the heat and are slow to cool. They are withstanding the firings, possibly because I painted the insides of them with a mixture of zircon flour and wallpaper paste. This coating, learnt during an inspection of a salt-glaze tile factory, does not prevent the build-up of salt and ash on the walls but hinders the fumes from penetrating through it to the bricks. The salt build-up on the walls helps in promoting an even salting in subsequent firings with the result that less salt is needed for these firings. Exploring wood firing possibilities further, in 1988 I built an anagama kiln, firing it for three or more days hoping to discover particular qualities of clay colour and the glaze effects possible with natural ash. In this kiln, small crucibles of salt are placed to enhance the ash vapours and to flash the sides of the work where the ash will not fall.'

(Above) Janet Mansfield, *Bowl*, stoneware, salt-glazed and fumed with stannous chloride during the cooling

Included in the research that Janet Mansfield has undertaken has been the use of sodium carbonate (soda ash) as an alternative to sodium chloride to form a vapour glaze. Another has been the replacement of part of the salt by borax. As one of the main attractions for her in the use of salt-glaze is in achieving the mottled orange-peel texture, these experiments have not been successful as yet. The soda ash glaze gives an uneven and greyed surface while the substitution of a 90% salt, 10% borax mixture gives a full gloss surface reminiscent of a limestone glaze. One advantage of the borax mixture is that satisfactorily melted glazes are possible at a lower temperature. Sawdust, sump oil, and colouring oxides can all be added to the salt mix for heightened temperature, oxygen reduction or colouring effects, and all offer more scope for experimentation.

During the cooling of the kiln, at temperatures between 900° and 1000°C, a process called *fuming* can give a metallic and lustrous surface to the salt-glaze. Volatile chlorides are thrown into the glowing kiln, and as they vaporise, they settle as a thin layer on the still-molten salt-glaze. Mansfield uses this technique, in particular with tin and iron chlorides, to obtain extra richness in the surface texture. 'All the ports and the dampers are opened to cause a draught of air to float the fumes through the

(*Above*) Heiner Balzar, *Relief*, salt-glazed stoneware, 1987

(*Left*) Heiner Balzar, vase, porcelain, 48 cm/h, 1989

(Above) Heiner Balzar, vase, salt-glaze over applied glaze

(Right) Robert Winokur, *Rome Alters Caesar*, stoneware, slips. ash glaze, salt-glaze, 30 x 25 x 25 cm

kiln,' she says. 'It is a matter of experiment to discover the right temperature and the right amount of chlorides to use. Too much chloride can result in dulled, creased cloth-like surfaces, while too little has no effect at all.'

In her work, Janet Mansfield seeks that combination of beauty and usefulness that gives pleasure to the user. Constantly seeing and learning is necessary for any artist, in particular a potter working in the somewhat unpredictable process of salt-glaze ceramics. She says: 'You have to keep looking at the work you are making; it should be relevant to the time and place in which you are living'.

Relevance to time and place is true for Heiner Balzar of Höhr Grenzhausen in Germany. Salt-glaze has always been part of his life, as it was of his parents' lives. His parents' workshop was instrumental in his development as an artist. His mother, Elfrieda Balzar-Kopp, was a major force in the studio pottery movement in the Westerwald area. Heiner Balzar, after studying with prominent potters and gaining his mastership, established his own workshop in 1967. It was here that he produced a prototype of a trolley kiln for salt-glazed ceramics, a technology which he has continued to develop. He has been influential in the Westerwald helping establish

(Left) Robert Winokur, *School of Fish*, stoneware, ash glaze, salt-glaze, 30 cm/h, 1988

organisations for the development of the arts and crafts and has won awards for his work and held many exhibitions throughout the world.

Balzar writes: 'My pots are fully completed on the wheel, not altered afterwards by tool or hand. The final form of the concept is already present during the throwing, with ideas of weight, balance and proportion determined. No further decoration is added as the glaze itself has the mission to support the form'.

Eckard Wagner describes Heiner Balzar's technical processes: 'He works with the light-burning clays of the Westerwald and porcelain. His glazes are based on felspar and coloured with metal oxides and ashes while the final character of the glaze is determined by the salt firing. Under reduction, the kiln is raised from 1200° to 1350°C, the salting taking place at 1260°C. An oxidation phase concludes the firing'.

The basis of the forms that Balzar makes is always the cylinder; this is often accentuated with ring-like sections. Wagner continues: 'Nothing is by way of chance, neither nuance or accent, in the vessel between the base and the opening. The various parts are calculated and there is precision of form. Yet the vessel seems a living being, growing or extending itself from the base. The intellectual aspect of the art is discernible in the traces of the hands. The unworked, unsmoothed throwing marks define, as a movement or motion, the static mass which one has to walk around in space. Balzar's vessels demand more than the optical experience; these active forms demand an active observer'.

In an essay on Balzar's work, Gerhard Gerkins says that Balzar is a self-sufficient artist who follows his own artistic intention. 'While his work has a crafts-like solidness, with its basis in the tradition of the Westerwald, it has no resonances of the past. He does not believe that craft work has a value only for itself, but says it must have an artistic basis which requires an ideological interpretation. He has remained true to salt-glaze and the vessel form yet works with an individual style. His vessels give the impression of great vitality of theme and a security in craft and technology.

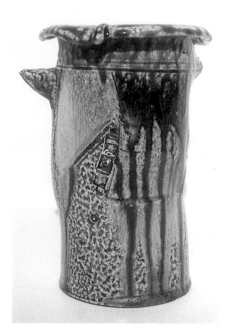

(*Above*) Robert Winokur, *The Fish Protrudes*, stoneware, slips, ash glaze, salt-glaze, 23 cm/h

(*Right*) Hans and Brigitte Börjeson, vase, slab-built stoneware, engobes, salt-glaze

They are heavy and mighty forms which he controls in a leisurely way. His well developed colour sense and his boldness and assuredness in the use of salt-glaze give a strong impression.'

Balzar is more concerned with what he makes than how he makes it. He is committed to making classic forms believing that the continuum of such forms is one of the tasks of the ceramist. It is natural for him to use the salt-glaze tradition of his family's workshop and from his Westerwald area, but he has taken it further, finding new ways to make 'not glass-cabinet pieces, but vessels that assert themselves freely in their own environment'.

Robert Winokur, a ceramist from Horsham, Pennsylvania, USA, sees the processes of the clay artist as akin to the forces of geology where, in some instances, these

forces are explosive, sculpturing the earth's surface, altering, folding, fracturing, carving and then covering the surface with volcanic and melted materials. 'Clay is a part of earth and the forces that affect the earth are no different than those that affect a lump of clay. The difference may just be a matter of scale. In ceramics, however, it is the artist who must make the decisions and influence the clay and firing processes; it is the artist who has the power to make creative judgments and evoke an aesthetic response.'

(Above) Hans and Brigitte Börjeson, *Step*, slab-built stoneware, salt-glaze

Winokur's salt-glazed vessels show the deliberate gestural movements of the potter's hands inside the clay forms and the molten layers of ash, glaze materials and salt on the external surfaces. 'I want my work to have the same abstract references as photographs taken of the earth from the air. Processes are to be used, not controlled, and fire is an element in the process and should be seen in the finished work. I want its unpredictability to be apparent. I see processes like throwing the clay, and then the firing, as a performance that one practices, rehearses, over and over again so that one may feel free enough and comfortable enough to improvise.'

The artist's career in ceramics has included numerous exhibitions, in particular with the Helen Drutt Gallery in Philadelphia. He has undertaken commissions, including a 180-piece salt-glaze tile panel in 1988, and his work has appeared in many publications. He received his Master of Fine Arts from Alfred University, New York, in 1958 after obtaining a graduate degree from Tyler School of Art and is currently Craft Department Chairman at Tyler. One of his concerns is the status of the ceramic vessel in relation to the fine arts.

He writes: 'Ceramic wares throughout history have been concerned mostly with the evolution of the container and the value of this art form has been judged in terms of the function it performed. Social and technological changes in the past twenty years have significantly eroded this tradition and ceramic objects of no particular function have become readily accepted, not as sculpture, though they are that, but as an art form with distinct characteristics. The traditional barriers based on media and technology, such as those that separated painting from sculpture, have also eroded so that one may easily see two-dimensional painted sculpture or three-dimensional

(*Above*) Christopher Staley

monochromatic painting. In this atmosphere I feel that forms once judged only in relationship to their function can now, without sacrificing their containership, come to be seen and accepted for their aesthetic merits alone. It's not a matter of making the perfect coffee cup. The cup is a form. What is crucial is what becomes of that cup when it is dealt with by someone who perceives the possibility of its becoming poetry'.

Winokur's belief that ceramics is an art form finds a positive response from all serious artists working in clay. Among these are Hans and Birgitte Börjeson, Danish salt-glaze potters who make functional ware and exhibition pieces. They also undertake commissions for architectural pieces such as fountains, archways, steps, basins, lampshades, tiles and free-standing sculptures. After studying ceramics in Sweden and Denmark they worked for three years with Harry Davis at the Crowan Pottery in England before they established their own pottery in an old school house at Soro in Denmark.

For the Börjesons, salt-glaze was a natural development, a wish to try out new ways. 'Right from the start of our pottery in 1963 we worked with the classical traditional stoneware glazes made from Swedish granite. This was a continuation of what we were doing at the Crowan Pottery, seeking out new materials, grinding them and testing them. The *tenmoku* and *celadon* glazes we made fitted our domestic pots well but the wish to do salt-glazing was always in our minds. An old kiln was acquired and we made our first experiments. The new vibrations really gave us a push forward and we worked day and night that summer. It was exciting to see how the shapes and designs changed to fit the new *expressions* we were achieving from the salt-glazing. At an exhibition in Copenhagen most of our pots were sold to colleagues, certainly the best response one can have.'

Since that time the Börjeson's work has become more sculptural, making architectural work for outdoors and indoors. The latest work is for a television studio for which they are making tiles and columns but, more than that, they are creating an atmosphere, helping to decide on colours, lighting and flooring in which their work will be featured. With increasing co-operation between architects and craftspeople they are optimistic about the future. Their work will all be salt-glazed, they say: 'Salt is a natural element; the colours, the texture, the whole atmosphere that salt-glazing entails makes it a natural element, and salt-glazed pieces fit in everywhere, inside as well as outside in nature. The challenge is the special designs needed to accentuate this ceramic process'.

American potter Christopher Staley is concerned with expressing his feelings and making them part of the rich tradition of pottery. 'I make pots for many reasons,' he says. 'The process suits me and the pots that I make can demand more of me than I expected. I have chosen to work in porcelain because of the range of vivid colours that are possible. Also, it has a sensuous quality for me. The firing process is critical to my work. It leaves its own mark which can be visually and conceptually evocative and in keeping with the unexpected quality which I look for in my work. I use copper as a colourant because it is unpredictable, sometimes green and sometimes red, depending on the atmosphere in the kiln. I often apply a thick slip, made up of my clay body, to the surface of a pot in a leather-hard state. The salt just adds alchemy.'

After receiving his Master of Fine Arts from Alfred University in 1980, Staley became a teacher and a practising potter with many exhibitions and honours to his name and is currently an artist-in-residence at the Archie Bray Foundation in Montana, USA. He is beginning to address social and environmental issues through his work and lectures. Staley says that one of the reasons he makes pots is because they are multi-levelled in terms of their audience. 'The relationship between myself

and clay is one of give and take. It is also a relationship of opposites, effortless-effort, illusion-reality, conscious-unconscious, and nature-machines. I'm striving for the most provocative meeting between myself and clay, the wheel and motion and also fire and surface, yet it's so elusive. In the end I hope my pots speak of strength through form, vitality through surface and a sense of purpose through function. I just remind myself that, if you struggle, it is a sign that you are alive.'

(*Above*) Christopher Staley, platter, porcelain, salt-glaze over copper and rutile glazes, cone 8, 58 cm/d, 1989

(*Above*) Ito Yushi, vase, white slip, salt-glaze

The full flavour of salt is demonstrated by these potters in the surfaces they achieve and the forms they make to display them. Japanese ceramist, Yushi Ito is a salt-glaze potter from Tokoname. He writes: 'In Tokoname, salt-glaze was popular and I heard that Shoji Hamada learnt salt-glaze techniques in this area. Today there are two or three kilns producing salt-glaze pottery in Tokoname'.

Yushi Ito learnt pottery from an early age as his father was also a potter. He worked in the pottery straight from school and when he was 23, he took the advice of his teacher to make salt-glaze pottery. Since that time 20 years ago he has been working with salt-glaze techniques, learning mostly by trial and error. His works have become

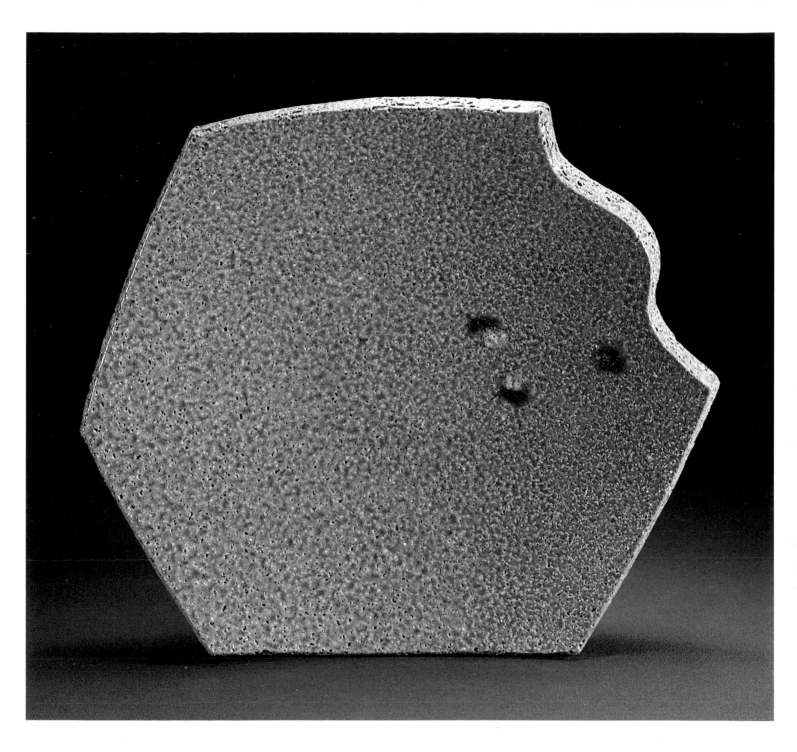

well known and in 1979 he was accepted as a member of the Nihon Kogei Icai (Japan Craft Association). He has won numerous medals and awards.

Using a white slip under the salt-glaze, wheel-thrown and handbuilt slab forms are decorated with symbols, the fine lines and superimposed colours giving the works a feeling of delicacy. Describing his philosophy, Yushi Ito said: 'I make pottery that reflects the ideas and emotions that flow through daily life'.

William Hunt, an American ceramist, has been intrigued with the full flavour of salt since his college days when he came across an old brown jar with 'a right-hand print showing five clear fingerprints in the slip on the side wall. As if frozen in time, you could see every whorl. I had never before noticed such a perfect imprint of a man from the previous century. Finding out about the potter and this jar became a low-grade obsession'.

(Above) William Hunt, footed platter, stoneware, perlite body additions, salt-glazed, 37 cm/d

As a part of this research Hunt discovered that 'pots were necessities to a community's development. For without pots, you couldn't reasonably ship or store, or preserve; metal and glassware were both expensive compared to clay. Pots could be stacked on barges plying the regional canal system'.

Eventually this one piece led to a collection of early American pots, mostly salt-glazed, and they, in turn, led to a desire to take a college course in art and pottery making. 'By graduate school I was hooked by pottery and also by collecting the salt-glazed works of the semi-anonymous potters from the US north-east. I realised that those plain, salt-glazed pots were unrecognised masterpieces. Most were quite equal to Brancusi's *Bird in Flight*, or the later work of the minimalists. Ever since, I've had a love for good form combined with undecorated ceramic surfaces.'

William Hunt studied many historical examples of salt-glaze including Bellamine jars and says: 'What I really found was my heritage. Salting is the only basic ceramic discovery which happened first in the western world and it came to America with the immigrants in the late 1700s and throughout the 1800s. I wrote my thesis on salt-glazing, starting by individually firing all material available in our glaze lab to see what each did alone. Then came combinations, until I understood empirically what happened in salt. Because I started out with very little knowledge, lots of fearlessness, time, and the school's equipment, I made many mistakes, but each was a learning tool'.

Hunt has had many exhibitions of his salt-glazed ceramics throughout the world. His writing and his editorship of *Ceramics Monthly*, are renowned and he has been invited internationally to speak on the ceramic arts. His personal commitment to salt-glaze continues. As he says: 'Salt gives a handsome skin through which the surface of the pot can be seen and I soon understood that whatever detailing goes into the kiln comes out sharper. Once you've smelled a salt kiln that is above 800°C, you never forget it. It quickens your pulse, sets a sense of anticipation, and starts dreams of something incredible happening in the firing. Once there is salt smoke in your nostrils, it never gets out of your senses. It starts a lifelong attraction to the next kiln load. Other methods of firing are nothing in comparison. Even wood smoke is like a steak without its salt, electric firing like an indoor barbecue'.

(*Above*) William Hunt, vase, stoneware,
salt-glazed, 14 cm/h

10.

A GENUINE CONTEMPORARY SENSIBILITY

T he enduring tradition of ceramic art has persisted over many millennia and seems sure to continue for many more. Today's ceramist works with clay and fire to articulate ideas that are relevant to contemporary society in all its fields: functional pottery, decorative vessels and sculpture. While some ceramists hearken back to the past, some express genuine contemporary values of the world in which we are living. Both are equally valid. For those ceramists working in the contemporary mode their practice uses the latest technology and their expression of art will give future archaeologists an insight into our values and technical expertise.

Rimas VisGirda, an American ceramist who was born in Lithuania, expresses through his work aspects of contemporary society, its material values, confidence and awkwardness and its brash confrontationalism. His work is contemporary in the idiom of the graphic narrative, looking to dazzle us with brightly coloured pictures that depict aspects of modern living. His subjects are urban; people, buildings, fashion and our culture are all subjects for experimentation and comment. He says: 'I think of myself as an observer. My work deals with the visual and conceptual elements in society that I find interesting, entertaining and consistent, with all the elements and details considered; the work should be purposeful and with a sense of unity. Conceptually, I intend the work to convey a sense of mystery, and an openness to different levels of interpretation that engage the viewer, provoke thought, and provide an opportunity for self-realisation about one's life, one's self, and society'.

VisGirda teaches ceramics at Illinois Wesleyan University, USA. He readily acknowledges the influence of Robert Arneson and the 1970s Funk movement both in his irreverent attitude and in his manipulation of clay. 'VisGirda enjoys evoking visual levity through the subject content and colours of his clay surfaces,' said Joan Mannheimer, writing as guest curator for an exhibition at the Olsen Larsen Galleries in 1984. 'In keeping with his irreverent attitude, VisGirda lets processes happen. He chooses a clay heavy with grog so that his stoneware pots have uneven, rough surfaces showing signs of warping, and blow-out holes made during the firing. He uses ordinary hardware bolts and screws to attach a rubber handle to a basket. Where angular pieces are added to a pot body, VisGirda lets the joining material, silicone, ooze from the seam in order to accentuate the process.'

'The two parameters I look at in any work,' VisGirda replies, 'are content and execution. Content refers to those qualities which open our mind and enable us to view or review our surroundings in a new light, to gain an insight into ourselves not possi-

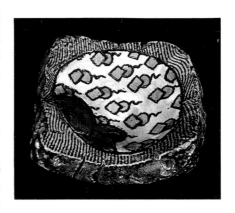

(*Above*) Rimas VisGirda, dish, salt-glazed stoneware, low-fire enamels, 17 cm/d

(*Facing page*) Monique Duplain-Juillerat, container, salt-glazed stoneware, coloured slips
Photograph: P.A. Nicole

(Right) Rimas VisGirda

(Facing page, left) Rimas VisGirda, *Woman with scarf*, salt-glazed stoneware, white slip, cone 9-11, low-fire lustre, cone 019–018, 70 cm/h, 1988. *(Right)* Rimas VisGirda, urn with pattern and lugs, salt-glazed stoneware, low-fire enamels, 40 cm/h, 1986

ble before such interaction. Execution refers to those qualities that have to do with the mastery of the media, good design, and attention to detail. Good work needs to allow different levels of interpretation; if the idea is too simplistic or direct, it becomes interesting or cute for the moment but, lacking thought or self discovery, is easily dismissed. Like good literature, a work should be subtle yet sophisticated, provoke rather than just entertain. The best work perhaps does both, it needs to touch our life, not just our eyes'.

Rimas VisGirda has been salt-glazing since 1969, and while he has always been unconventional in wishing to add low-fire coloured enamels to the surface of his pieces, he still seeks the depth and richness of the surface and hardness that lies under the transparent colour of high-fire salt-glaze. For him, the salt-glaze is a perfect background foil to his colourful imagery. With the intention of combining sculpture, drawing and ceramics, he makes forms that are slab-built or extruded, often cylindrical, with flat planes suitable for the imposed images. He uses a white slip to cover the clay, draws on the surface with a pencil, then covers the entire piece with a wax resist. Using a sgraffito tool, he scratches through the wax, back through the drawing to the clay body. These lines are then filled with a black slip to define their outlines. After salt-firing to cone 10, lustres and underglaze colours are applied and a further firing to cone 019 is given. Sometimes many low-temperature firings are needed to complete all the colours required for the image.

An increasingly popular and contemporary use of salt in firing is at low temperatures. American ceramist Paul Soldner, an early exponent of this technique, used it as part of his repertoire of *raku* ceramics. David Miller, the English ceramist now living in France, has developed this technique, combining various colours with the matt texture of low-fired clay. 'There exists a close association between the concep-

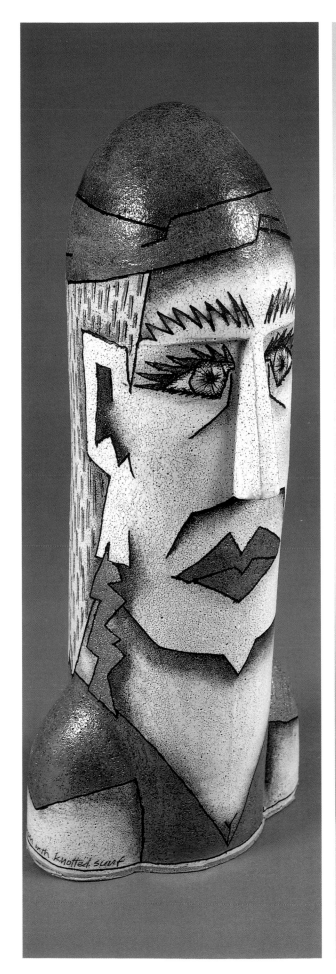

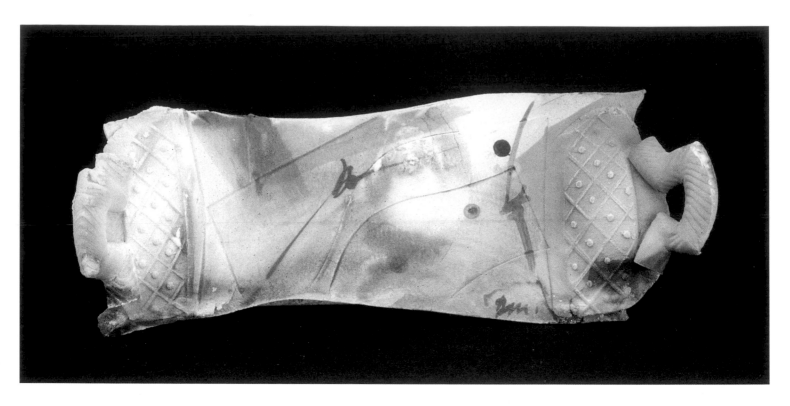

(Above) David Miller, plate with rolled handles, low-salt firing, 45 cm/w

tion and realisation of a piece using this technique,' he says. 'Marks or incisions made in the soft clay remain clearly defined in the dry surface, giving an aspect of freshly worked clay. Ideas can evolve for me with this process.'

Miller explains this technique: 'The function of salt, at a temperature below that at which a glaze is formed, is to draw out subtle nuances of colour in both the clay body and the slip. In order to arrest the formation of a glaze, the temperature must not exceed 1000°C. The way in which the kiln is set has an important influence on the distribution of heat and the circulation of the salt vapour. The pieces can be stacked without kiln shelves, but they should be protected from direct exposure to the flame by a bagwall. The flame picks up salt which has been placed in bowls and packed amongst the pieces along with other combustible materials, and carries the vapour through the maze of pots. The firing normally lasts between three and four hours, and after a fairly rapid temperature rise, in a neutral atmosphere, 700°C is reached.

'When the salt first shows signs of vaporising and a clean flame is seen winding or passing through the stacked ware, the firing is slowed down to give the salt time to work. Any direct interaction between the salt and the pots will tend to leach out the iron in the clay, often accompanied by a watermark effect due to the running of the liquefied salt. Pink flashings, caused through the use of small amounts of copper in the slip, are more easily developed by a passing vapour rather than through direct contact with the salt. These markings can be further encouraged by mixing copper sulphate with the salt and also by introduction of wood, sawdust, vermiculite and charcoal amongst the pots; these organic materials burn away causing a localised reduction to occur. Bold brush strokes can be affected drastically by the caustic attack of the salt, softening and diffusing the sharp edges and lines.'

While David Miller feels that he is working within the pottery tradition, his pots are contemporary interpretations of vessel forms, with the functional aspect suppressed in favour of other references. He says: 'I find throwing a way of developing and carrying through an idea, linking the making process with a mixture of ideas and feelings which together stimulate a creative flow. I usually start by throwing a rea-

sonable number, say 10–20, of geometrically based forms which are remodelled at a later stage. At the leather-hard stage, foot-rings, knobs etc. are added. Each piece develops an individual character which I try to respond to but, in the end, it is the firing process which takes command, modifying previous intentions. The pots have simple uniform beginnings but become asymmetric through a process of reforming and by the placement of their attachments'.

Also looking to express contemporary qualities of shape and colour but in high-fire salt-glaze is Dutch potter, Paulien Ploeger. She trained in ceramics in Holland and England and then established her studio in St Jacobi-parochie in the Netherlands in 1981. She has exhibited in Germany and Holland and her work is documented in various catalogues and magazines. She takes what she needs from past methodology and traditional salt-glaze techniques and strives to use them in new ways. She writes about her work: 'Ten years ago I took up salt-glazing. This technique comes very close to that directness and intensity I experience when I am throwing pots: wet and plastic. Everything I do during the making of a pot is emphasised by the salt and the fire. For me it is a challenge to combine the traditional techniques, such as throwing and salt-glaze, and transform them into a contemporary mode'.

Inspired by jazz music, Ploeger searches for similar harmony, discord and improvisation in her work. This results in angular forms which challenge equilibrium and on which the decoration even further emphasises the form. Describing her technique she says: 'These pots are built up from several thrown pieces cut at different angles, turned around and placed out of centre to become new asymmetrical pieces. I try to incorporate my use of colour with the shape. The techniques I use are paper and wax resist. Pots are dipped and painted with slips that I colour with oxides and stains. I sometimes combine straight lines with freely trailed lines to get a sense of movement

(Above) David Miller, cone teapots, coloured slips and oxide decoration, low-salt firing, 33 cm/h

(Right) Paulien Ploeger, square dish, salt-glazed stoneware, 35 cm/w

and action. These pieces show the discipline of the wheel and at the same time they move away from that in a contemporary and playful fashion'.

Owen Rye is an Australian ceramist who has a contemporary attitude to ceramic art, both in aesthetic content and the use of materials. With a PhD in ceramic technology from the University of New South Wales, he is able to combine the latest technical advances in clay body formation with traditional values of making pots. While his main interest is in wood firing, particularly in anagama-type kilns, which fire over periods of three to six days, he can see parallels between this type of firing and his work with salt-glazing, especially when salt-glazing in a wood-fired kiln.

'I have been experimenting with salt-glazing for ten years,' he writes, 'initially by placing pots and salt in saggars and firing them along with regular glazed stoneware. More recently I have worked with porcelain bodies to produce blue and white wares which sit companionably with a long-standing interest in porcelain. Both wood and salt types of firing require carefully considered kiln packing, and the setting determines the nature of the surface, particularly flashing and the distribution of marking. In both types of firing the surface coating, salt in one, ash in the other, is produced by the live flame carrying the active substances around the ware. Both also require a sensitive consideration of surface markings during the making. Plain, smooth surfaces, textures, and lines, as well as the fine or coarse nature of the clay itself, all contribute to the final character of the work. Salt-glazing is unparalleled among ceramic processes for preserving the immediacy and spontaneity of the making and it is this quality which leads me to bouts of salting, as a contrast to the slow, considered nature of anagama firing'.

Gerd Knäpper is a German ceramist who has spent most of his working life as a potter in Japan. One valuable aspect of his work is his ability to take influences from two countries and turn them into a contemporary expression of ceramic art which is

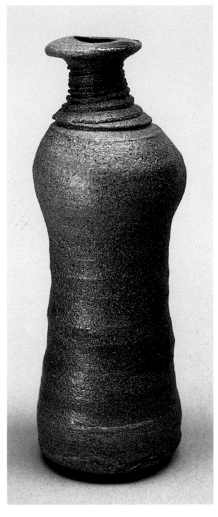

(*Above*) Owen Rye, vessel, fired with charcoal in salt kiln, 1989

(*Left*) Paulien Ploeger, *No reeds*, salt-glazed stoneware, 35 cm/h

intercultural. It is in Japan that he has achieved his greatest success, having won many prestigious awards for his work. He writes: 'It seems strange for a German to discover the beauty of salt-glaze and its firing process in Japan. Compared to Germany, where salt-glaze firing probably started around the 13th century, the history of salt-glaze in Japan dates back only to the Meiji Restoration time between 1868 to 1912. Salt-glaze firing techniques were first introduced to Japan from Germany, with the introduction of production methods to produce stoneware sewer pipes and chemical containers. It seems that this method of firing to produce durable and acid resistant ware was unknown in Japan previously. One of the reasons for this might be the non-existence of large, natural salt deposits.

(Right) Gerd Knapper

'However, in early days, perhaps unknowingly, a flair of sodium changed the char-acter of some ceramic wares at Bizen. The potters located close to the ocean used to stack their fire wood to dry along the shore line, where winds entering from the ocean carried and deposited salt on it; pots were accented by the sodium fumes pre-sent in the fire wood. During further development, pots were stacked with rice straw in between them, leaving reddish marks on the surface contacted; this became one of the adopted techniques to decorate traditional Bizen Pottery. Today, rice straw or rice straw rope is soaked in a strong salt water solution, and then wrapped around a vessel before being fired in an oxidised atmosphere to give beautiful flashings on the surface of the pot where the sodium from the salt combines with the silica in the clay.

'I choose salt-glaze firing particularly for the finishing of tea ceremony bowls, looking to combine natural beauty in imperfect shapes, and the impurity of this unpredictable glaze. For me, salt-glaze firing is not an everyday process but a stimu-lating experience, full of excitement.'

Seiro Maekawa, Director of the National Museum of Western Art in Tokyo, wrote in the book *Gerd Knäpper*, published by Kodansha in 1989: 'Gerd Knäpper's creations are more Japanese than the works of many Japanese. And still the origins of his art remain undoubtedly German, as expressed clearly in the massive feeling of its shapes. Interestingly, he uses the Japanese wave motif. Just like 300 years ago, when porcelain and motifs from the town of Arita found their way across the ocean to Germany, where they still flourish in Meissen, may the bold and at the same time refined wave motif of Japan cross the ocean to Germany and be transplanted there by the hand of Gerd Knäpper. It is not without symbolism that this motif is called "blue waves of the great ocean" '.

(Left) Gerd Knapper, *Venus*, hand moulded, 46 cm/h, 1985

Gerhard Bott, Director of the German National Museum, writing for the same publication said: 'Knäpper's ceramic works are clearly identifiable. They differ from contemporary Japanese wares by the unique shape and also by the motif. The German potter in Japan feels less bound by the traditional and utility-oriented basic shape of Japanese ceramics. In addition, he invented new forms and new methods of decoration. The bold and individualistic way he decorates his porcelain work with a delicate line of inlaid colour is a form of expression not represented in the repertoire of either his Japanese or European colleagues. Knäpper also constructs his ceramics into independent structures which, both in respect to shape as well as the structure of its surface, can make the claim to be sculptures. The creations are not called vases by him, but *Torso* and *Venus*. Gerd Knäpper has succeeded in becoming a link between two different cultures'.

John Neely, an American ceramist, also takes a multicultural approach to his work. He makes contemporary and unique container forms, teapots and vases which have strong clean profiles with contrasting freely formed attachments. Rather than a veneer, the surface is deep and matt and seems to be part of the clay body itself. His pieces are the result of considerable personal inquiry; his research has taken him to Japan where he spent many years studying and working as a potter. Neely has been firing with salt for twenty years and although it is not the only technique that he employs, it accounts for a significant part of his repertoire. 'Salt seems to be an indispensable part of an equation,' he says. 'Information gleaned by firing with salt has enabled me to understand other processes, and salt, in turn, has been informed by the exploration of other techniques. I have been firing with wood for nearly as long as I have with salt, and although I have seldom felt the urge to combine the two, it

(Above) John Neely, basket, salt-fired stoneware, 35 cm/d, 1988

seems that they have been complementary in my growing understanding of ceramics.'

One of the problems that has intrigued him in his research both in Japan and the US is the differing effects possible on clay and glaze during the cooling cycle of the kiln. He writes: 'I have long been intrigued by the flashes of colour, of pink, red, and orange, that occur in both the wood kiln and the salt kiln. A few years ago, the school where I am employed, Utah State University, funded the construction of a new salt kiln to investigate this phenomenon. Having noticed that flashing is entirely absent in oxidation firings, and also not to be found in work subjected to reduction cooling, I started from the premise that this colour is a re-oxidation phenomenon. I built a hard brick kiln and insulated it with soft brick and fiber, the object being slow cooling. As the red colour was the result of haematite development, I reasoned that an extended cooling would allow greater crystalline development'.

Neely continues: 'I knew before going into the project that cooling was not the only factor involved. While working at the Kyoto University of Fine Arts in Japan, I built a salt kiln that allowed me to compare results with a variety of clays fired in both that kiln and the anagama. The native kaolinitic clays of Shigaraki, those called *Kinose* and *Shinohara*, which show such remarkable colour in the anagama, also exhibited this tendency in the salt kiln. Noting that the greatest colour development was in spots where glaze build-up was light, rather than using less salt, which would result in unacceptably dry surfaces, I decided to adjust the body in such a way

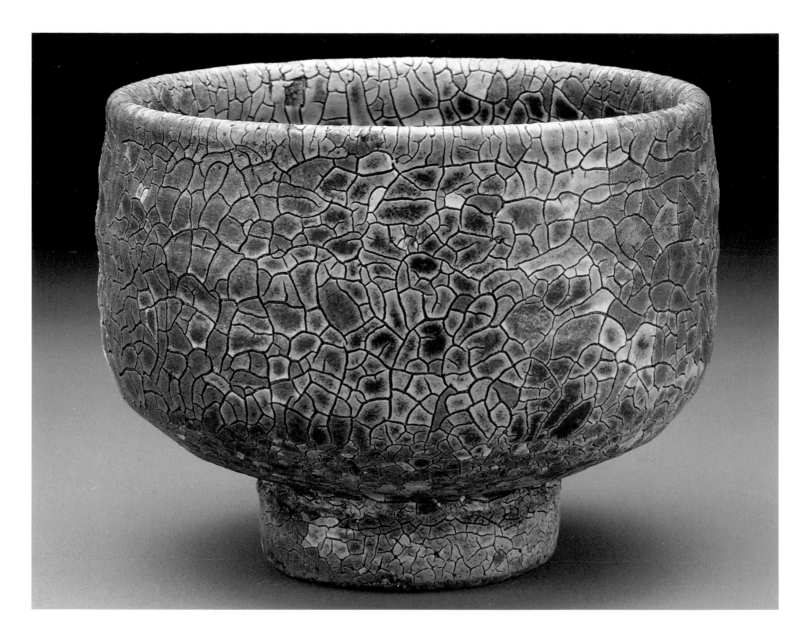

as to inhibit glaze build-up. I also used these combinations as slips. The salt kiln in the USA allowed me to continue this line of investigation. Where previously the object had been economy, an acceptable glaze coat at the lowest possible price, my goal was an aesthetic one.

(Above) John Neely, bowl, salt-fired stoneware, 10 cm/h, 1989

'I wanted the colours mentioned in the industrial and scientific literature only as faults and defects. Mineralogical factors and particle size are often critical in body formulation. Focused as I was on surface effects it was some time before I managed to develop bodies that fell within the limits I had established and met the demands of workability. We speak of salt-glaze as if it were a glassy coating similar to other glazes. In fact, it is more like what is usually called the clay/glaze interface without the glaze. During salting, alkaline vapour eats away at the surface of the ware, dissolving the body into glass. This glass, in turn, while incorporating more and more sodium at the surface exposed to the kiln atmosphere, is dissolving more and more of the body.'

John Neely has experimented with using magnesium and calcium compounds, both of which are reported to inhibit salt-glaze formation. 'My experience has been rather that magnesium inhibits only the development of a glossy glaze coat; the addition of talc, a magnesium silicate mineral, can yield satin-matt salt-glazed surfaces of great potential and are easily achieved. Even more interesting are surfaces resulting

(*Above*) John Neely

from the addition of calcium. By adding wollastonite, a calcium silicate mineral, not only are the working properties of the clay enhanced but, with sufficient salt, incomparable crystalline matt surfaces are possible.'

Many of his slip recipes are based on a mixture of two kaolins plus 17 silica 200# and 23 felspar to give a salmon colour, or a mixture of three clays and 12 wollastonite and 5 felspar for a matt yellow. A gloss trailing slip is made from 50 ball clay, 50 nepheline syenite; this responds well to colourants and is not limited to trailing.

The scope of Neely's research has not been confined to surface and texture. His forms, while based on traditional container shapes, have a sharpness of edge and concept which makes them fully contemporary. His ceramic pieces are at home with contemporary architecture, fulfilling a modern use for modern times. The recipient of a 1989 fellowship for the Visual Arts, he states in the accompanying catalogue: 'I quite consciously allow material and technique to generate form and am most satisfied when these forms find some resonance with the demands of utility'. In an article for *Ceramics Monthly*, April 1988, titled 'Nice Cooling', he wrote: 'All the technology available doesn't seem to make it easier to produce really good pots. Throughout history it seems that good potters have managed to make good pots regardless of the available technology. It does seem significant, however, that they keep pushing the limits of that technology, striving to understand the mechanism of their craft. I see many of today's potters, myself included, engaged in a similar struggle'.

A contemporary expression in ceramics is the depiction of architectural detail, of using a fragment to reflect on civilisations past or taking the idea of a container to symbolise the wider context of home and environment. In this category is the work of Monique Duplain-Juillerat, a Swiss ceramist from Allaman. Salt-glaze combined with colour is used by her to portray her *architectural dreams*. 'They are an amalgam,' she says, 'of my relationship to the Orient and the Occident, trying to emphasise archaism and refinement, contrasts in a wide sense of inside and outside, dark and light, shiny and matt, severe and playful'.

Duplain-Juillerat trained at the Ecole Suisse de Ceramique, Vevey, and Ecole des arts décoratifs, Geneva. After working in Switzerland, France and England, she took time to study in Korea and Japan before setting up her own studio in Allaman, Switzerland. She has had numerous exhibitions, contributed to international group shows, and won distinctions for her work.

She writes: 'The salt reacts like a catalyst with the stoneware, the slips and the glazes; their interaction becomes a decoration. Something unforeseen happens, partly provoked but always a surprise. The forms are mostly thrown as I have always been fascinated by centrifugal power. The throwing gives a primary shape, like a sketch, and further deformation and modelling and texture gives me the final form. For decoration I use patterns inspired by my environment, impressions of textiles and woodwork. I use vitrifying slips, either trailed or using wax-resist, and combine these with ash glazes as well as felspathic, magnesia or titanium glazes. I fire my kiln to 1290° or 1300°C in reduction. In wood-fired kilns there is an interaction between clay, ash and fire. With salt I can re-create this quality'.

Stefan Emmelmann has his ceramic studio in Austria. He studied in Germany and England and has given workshops and held exhibitions throughout the world. Looking now to make large architectural work, stoves, sculptures, fountains and murals, he is using salt-glaze effects in various ways to impart the surfaces for his work. He writes: 'My initiation to salt-glazing was in 1977. I had returned from a two-year working visit to British studios and was thoroughly steeped in the Leach tradition. Nevertheless, I wanted to gain some experience with typically German pottery and found a job in a workshop in the Westerwald area where the well-known

(Left) John Neely, *Cruet*, salt-fired
stoneware, 15 cm/h, 1988

blue and grey salt-glazed pots were still being produced in the traditional way. The
sheer size of the operations was impressive, especially their huge wood-fired kilns. It
was definitely a decisive experience. In this workshop I met my future wife, Reinhild
Frech, and together we decided that wood-fired salt-glazed stoneware was the only
real one for us although we favoured brown hues for our work.

'In 1979 we established our own studio in Austria and produced functional pottery with this technique for several years, using local clays with an iron content of
4–5%. These clays yielded a wide range of rich brown colours which contrasted
beautifully with cobalt decoration. For our 3 m³ (100 cu.ft) salt kiln we poured about
18 L (4 gal) of salt in two or three charges at top temperature on to the wood in the
fireboxes. However, I began to feel increasingly uneasy about our contribution to
pollution once a month and, at the same time, we wanted more colour in our pottery.

(*Right*) Stefan Emmelmann, tile, low-fire salt

I also began to feel that our production pottery hadn't left any room for the development of new ideas. So my work slowly developed towards large-scale architectural objects, which I approach sculpturally, with an emphasis on surface textures.

'When firing these new pieces, we place small amounts of salt in little cups between the pieces in the kiln setting; this gives us some beautiful flashing on clay and slips. The pigments and glazes sometimes vary dramatically, marking the passage of the long wood flames carrying the fumes through the kiln. I am thoroughly convinced that this subtle use of salt can give us just as exciting results as its regular use.'

Stefan Emmelmann's concern for the polluting aspects of salt-glaze is a real one. It is important for all ceramists using the technique of salt-glaze to realise the potential of salt fumes to corrode galvanised iron and burn the lungs if inhaled. The chlorine coming from the kiln chimney combines with the water vapour, in the air and the combustion process, to produce a dilute form of hydrochloric acid. Steps should be taken to minimise any harmful effects of these fumes. Commonsense and practical economy should ensure that salt is thrown into the kiln only in amounts that will provide the desired effect. More is unnecessary. The kiln should be well sealed so that all the fumes go through the flue system and the area around the kiln should be well ventilated. The kiln should not be charged with salt until the clay body of the ware has reached a state of vitrification. Then, all the salt in the kiln will be effective, working on the surface of the pots instead of going straight out to the atmosphere.

Max Murray of the Chisholm Institute in Melbourne, Australia, has developed a fume washing system to attach to the chimney of a salt-glaze kiln. Writing about this system for *Pottery in Australia*, Vol.27/1, he describes how the gases produced during the salting process are directed, by means of a secondary flue, into a stainless steel washing tower which has a swirl-jet spray head attached near its highest point. He writes: 'The gases are forced to pass into the water spray where most of the acid and

(*Above*) Stefan Emmelmann, tile, low-fire salt

chlorine are washed out. This also has a secondary benefit of lowering the temperature of the remaining exhaust gas and also helps to damper the kiln and maintain reduction at the critical moment of salting. Washing is only carried out during the salting period and the deflecting plate is open for the rest of the firing'. The amount and colour of the fumes is different when a washing tower is used. The acid run-off water from the tower is neutralised by running it over limestone chips. If ceramists using salt-glaze techniques are careful, as they should be with every aspect of the practical application of ceramic technology (that is, the use of chemicals and fire), there should be no danger. Heat from the kiln's chimney will damage trees but the vapour from the salt fumes will not.

There can be no conclusion to this book. More and more ceramists throughout the world are continuing to discover possibilities in the salt-glaze technique. The spontaneity and directness of the technique, both seen to be essentially desirable qualities of current aesthetic attitudes, attract many artists. Ceramics is a classic and timeless art. We can learn of past civilizations from studying pottery and we can be excited about the future. Salt-glaze will be a part of that future. Working with clay and glaze and firing with kilns gives an artist a means to express personal creativity and aesthetic values. In turn, these ideas are modified by the beliefs, social conditions and practices of the artist to provide the differences that make ceramics a living art form. The ceramists who have contributed to this book have generously explained their approach to their work, from both their aesthetic and technical viewpoints. The reader is thus invited to share the artists' dreams and their practice and love of salt-glazed ceramics.

Walter Keeler, jug, 18 cm/h, 1989

INDEX AND GLOSSARY

FURTHER READING

Books and Periodicals:

A Collector's Guide to Modern Australian Ceramics by Janet Mansfield, Craftsman House, Sydney, 1988

Ceramic Science for the Potter by H.G. Lawrence, Chilton Press, Philadelphia, 1973

Salt-Glaze by Peter Starkey, Skill Books, Pitman, London, 1977

Salt-Glaze Ceramics by Jack Troy, Watson-Guptill Publications, New York, 1977

Ceramics: Art and Perception (periodical), Paddington, Sydney, Australia

Monographs and Pamphlets:

La Borne, A Potter's Village, Association des potiers de La Borne

Potiers d'aujourd'hui au pays de La Borne by Robert Chaton

Salt-Glazing by Don Reitz and others, American Crafts Council, 1972

Numerous texts on the history of salt-glaze have been written. Research into German salt-glaze could be undertaken through the Westerwald Museum in Höhr Grenzhausen or the Hetjens-Museum in Dusseldörf. In the USA, Elaine Levin's *The History of American Ceramics*, Abrams, New York, 1988, gives a good overview and Emanuel Cooper's *A History of World Pottery*, Batsford, is a worthwhile reference.

In Australia, there are a number of specialist books: *The Lithgow Pottery* by Ian Evans (Flannel Flower Press), *Bendigo Pottery* by Paul Scholes (Lowden), *Australian Studio Pottery and China Painting* by Peter Timms (Oxford University Press) and *South Australian Ceramics* by Noris Ioannou (Wakefield Press).